Lecture Notes in Mathematics

Edited by A. Dold and B. Eckmann

1083

Daniel Bump

Automorphic Forms on GL $(3,\mathbb{R})$

Springer-Verlag
Berlin Heidelberg New York Tokyo 1984

Author

Daniel Bump
Department of Mathematics, University of Texas at Austin
Austin, TX 78712, USA

AMS Subject Classification (1980): 10D20, 10G05, 10D24, 43A80

ISBN 3-540-13864-1 Springer-Verlag Berlin Heidelberg New York Tokyo
ISBN 0-387-13864-1 Springer-Verlag New York Heidelberg Berlin Tokyo

Printed in Germany

Printing and binding: Beltz Offsetdruck, Hemsbach/Bergstr.
2146/3140-543210

PREFACE

The theory of automorphic forms is one of the most satisfying and beautiful branches of mathematics. Automorphic forms may be defined arithmetically on any reductive Lie group, and these have been studied intensively for many years. The most familiar ambient groups are GL(2) and the symplectic groups, which attracted attention because of their relation to the moduli problem for elliptic curves and abelian varieties.

In part because of Langlands' conjectured principle of functoriality, another class of reductive groups, namely, the general linear groups, has emerged as the focus of recent intensive research. The general linear groups will inevitably play a central role in the theory of the L-series associated with an automorphic form.

Before studying automorphic forms on GL(n), it is natural to consider the simplest, most familiar case, namely, that of automorphic forms on GL(2). However, in this simplest case, many features of the general situation are not apparent. One looks next to GL(3).

The theory of automorphic forms on GL(3) was greatly advanced by the work of Jacquet, Piatetski-Shapiro and Shalika [15], who proved the converse theorem in the Hecke theory, and in the process, developed much important machinery. Their work, as that of most experts, uses the full machinery of representation theory.

In these notes, we shall attempt to lay a reasonable foundation, ›m a classical point of view, for the study of automorphic forms on ,$\mathbb{R}$). Here "classical" means that we shall consider the forms as ›d on the real group, rather than the adele group, and that we ›void, as much as possible, the language of representation theory.

To a certain limited extent, we shall employ the language of re[illegible] tions to make precise formulations, or to explain connections with the literature.

We hope that the present notes will be found accessible. We believe that we have covered the most important basic topics on GL(3) as simply as we are able. In the early chapters, we have attempted to aid readability by stating the results of the chapter before the proofs -- the reader may defer the proofs to a second reading if desired. There is considerable independence between the early chapters. Let us point out that although Chapter II contains some rather long and tedious computations, the material in this chapter, while important, is largely unneeded for the sequel. Chapter III is essential but may be deferred. The reader wishing a quick entrance to this book might read Chapter I, then start with Chapter IV, going back to Chapters II and III when necessary.

The automorphic forms which we shall consider on GL(3) are those corresponding to the principal series representations of $GL(3,\mathbb{R})$. These are analogous to the forms on GL(2) defined by Maass [22]. We shall restrict ourselves to forms which are automorphic with respect to the full modular group $GL(3,\mathbb{Z})$.

The main topics which we shall consider are the theory of Whittaker functions, their differential equations, and their analytic continuation and functional equations; Fourier expansions on GL(3); the Fourier expansions of the Eisenstein series, and the theory of Ramanujan sums on GL(3), which arise in the Fourier expansions; the analytic continuation and functional equations of the Eisenstein series; the polar divisor of the Eisenstein series; the interpretation of the Fourier coefficients of the Eisenstein series as generalized divisor sums, expressed in terms of Schur polynomials; the L-series associated with an automorphic form, their analytic continuation and functional equations; Hecke operators, and the Euler product satisfied by the L-series associated with an morphic form; double L-series; and the double Mellin transform of Whittaker functions.

Some of the results are new, but the general outline of this theory been known to experts for several years now. The biggest novelty of our approach is that it is very direct and explicit.

In Chapter I, we introduce the topic by considering automorphic forms on GL(2) from the same point of view that we will take for GL(3). We assume some familiarity with this theory.

In Chapter II, we consider the differential equations satisfied by the Whittaker functions, those special functions which are the basic building blocks of automorphic forms. We will see that the Whittaker functions satisfy an overdetermined system of differential equations, the dimension of whose solution space is six. Only one solution satisfies a growth condition. This uniqueness is a special case of Shalika's local multiplicity one theorem [28] which, unfortunately, we were unable to prove along the lines of this chapter. We will exhibit six linearly independent solutions of the Whittaker differential equations as generalized hypergeometric series, and one solution which satisfies the growth condition, the latter as either a Mellin-Barnes integral, or as a linear combination of the six hypergeometric Whittaker functions. From the latter Mellin-Barnes integral, we will prove that the Whittaker function is of rapid decay, a fact which is needed for convergence proofs in subsequent chapters.

In Chapter III, we shall consider Whittaker functions which are defined as definite integrals, after Jacquet. Jacquet observed that these Whittaker functions had analytic continuation and functional equaions, as may be deduced from the corresponding properties of the Eisenein series, and set out to give direct proofs of these facts. We all follow his method (based on Hartogs' theorem). The context of cquet's results is very general -- we will be concerned with only a ry special case. It follows from the local multiplicity one theorem that Jacquet's Whittaker function is equal to the Mellin-Barnes of Chapter II, up to constant multiple. In Chapter X, it will b that these functions are actually identical. We will also consi degenerate Whittaker functions required for the Fourier expansion the Eisenstein series.

In Chapter IV, we show that an automorphic form has a Fourier expansion, the coefficients forming a two-dimensional array. On the adele group, the analogous result is a well-known formula, due independently to Piatetski-Shapiro [24] and Shalika [28], reconstructing the form from its global Whittaker function.

Chapters V, VI and VII are devoted to the Eisenstein series. The Fourier expansions of the Eisenstein series on GL(3) were obtained independently by Bump, by Vinogradov and Takhtadzhyan [37], and by Imai and Terras [11]. The analytic continuations and functional equations of the Eisenstein series (on a general reductive group) were obtained by Langlands [20].

We will only consider the "minimal parabolic" Eisenstein series -- the most important sort. It is also possible to build Eisenstein series by inducing GL(2) cusp forms up to GL(3). For the latter "maximal parabolic" Eisenstein series, we refer to Imai and Terras [11].

The Eisenstein series are defined initially as sums over the orbits of $\Gamma_\infty\backslash\Gamma$. In Chapter V, we will determine these orbits as follows. First we consider the orbit space $G_\infty\backslash G$. We map this space bijectively onto a certain hypersurface in Euclidean space, by means of Grassmann coordinates. Then, we give a criterion for an orbit to contain an integral orbit. This amounts to a determination of $\Gamma_\infty\backslash\Gamma$. We shall relate our theorems to the Bruhat decomposition, by giving the parametrizatio explicitly on each Bruhat cell. It is interesting to note that our approach is very close to that of Vinogradov and Takhtadzhyan [37] (c their Lemma 1).

Chapter VI is devoted to the theory of Ramanujan sums on GL(3). These are exponential sums which occur in the Fourier expansions of the Eisenstein series. The corresponding sums on GL(2) were introduced by Ramanujan [27], and are known classically as Ramanujan sums; we have retained the same name for the GL(3) sums. We will evaluate these sums in terms of the "Schur polynomials," which are the characters of the finite-dimensional irreducible representations of GL(3). The occurrence of these characters in the Fourier coefficients of automorphic forms is of supreme importance. We have attempted to bring out this aspect systematically in the last chapters of the book.

Chapter VII includes the Fourier expansions of the Eisenstein series, and the analytic continuations and functional equations found by Langlands. The Fourier coefficients are generalized divisor sums, defined by means of the Schur polynomials. Also in this chapter is a discussion of the polar divisor of the Eisenstein series.

Chapter VIII defines the L-series associated with an automorphic form, and shows, following Jacquet, Piatetski-Shapiro and Shalika, that this L-series has an analytic continuation and functional equation.

Chapter IX is devoted to the Hecke operators, and the properties of Hecke eigenforms. It is shown that the L-series of a Hecke eigenform has an Euler product. Also considered are the double L-series associated with an automorphic form.

Chapter X is devoted to the computation of the double Mellin-transform of the Whittaker functions. This chapter completes the proof that Jacquet's Whittaker function is given by the Mellin-Barnes integral of Chapter II.

In the matter of acknowledgements, I have discussed the matters of this work with many people, and it would be hard to name all of them. A previous, unpublished but widely circulated version of this work included my dissertation at the University of Chicago. I would like to thank my advisor, Professor Walter Baily, and Dinakar Ramakrishnan, for insisting on many needed changes in that manuscript. I am particularly indebted to Walter Baily, Joe Buhler, Dorian Goldfeld and Audrey Terras, for their interest and encouragement kept me going through a difficult time. I would like to thank my parents, for making it possible for me to work undisturbed during the preparation of that first manuscript. I would like to thank Jerry Tunnell for his encouragement during that time also, and I would like to thank Professor Paul Sally for refereeing my dissertation.

Since that earlier version of this work, I have learned much which has resulted in the improvement of this work. I would like to thank Jeff Hoffstein, Dorian Goldfeld and Solomon Friedberg for what I have learned from them. I would like to thank Springer Verlag for their efficiency and cooperation in producing this work. And, I would especially like to thank Jan Duffy for her accurate, efficient, and cheerful production of a difficult manuscript.

Many computations in this work were too difficult to do conveniently by hand, especially in Chapter II. For these, I used muMath, a computer algebra system for microcomputers, which I modified to do computation in the noncommutative algebra of differential operators. muMath is a very slick, useful product, which I do not hesitate to recommend. So, I would like to thank the Soft Warehouse, P.O. Box 11174, Honolulu, Hawaii 96828, for their contribution to this work.

I would happily be apprised of any errors which anyone should find in this work.

May 31, 1984
Department of Mathematics
The University of Texas at Austin
Austin, TX 78712

FREQUENTLY USED SYMBOLS AND NOTATION

Although a reading of Chapter II may be deferred without loss of continuity, it should be pointed out that many notations which are used constantly throughout the text are introduced in the first four pages 17-20 of that chapter. These should be assimilated.

Only notations which are used in more than one chapter are enumerated here. Thus, if you cannot find a symbol or notation in this table, look for it in the chapter you are reading.

Page		
65	a_{n_1,n_2}	Fourier coefficients
73,74,82	A_1,B_1,C_1,A_2,B_2,C_2	Invariants of $g \in G_\infty\backslash G$
5	$e(x)$	Exponential function
	G	$GL(n,\mathbb{R})$, $n = 2$ or 3
3	G_ν	GL(2) Eisenstein series
101	$G_{(\nu_1,\nu_2)}(\tau)$	GL(3) Eisenstein series
2,17	$\mathcal{H}$	Homogeneous space
5	$I_\nu^*(\tau)$	
19	$I_{(\nu_1,\nu_2)}(\tau)$	
17	K	O(3)
49	K_ν	Bessel function
13,133	$L(s,\phi)$	L-series
83	$R_{A_1,A_2}(n_1,n_2)$	Ramanujan sum
86	$r_{A_1,A_2}(n_1,n_2)$	Ramanujan sum
87	$s_{A_1,A_2}(n_1,n_2)$	Ramanujan sum

FREQUENTLY USED SYMBOLS
AND NOTATION (CONTINUED)

Page		
83	$S_k(\alpha,\beta)$	Schur polynomial
84	$S_{k_1,k_2}(\alpha,\beta,\gamma)$	Schur polynomial
18,19	W	Weyl group
18,19	w_0,w_1,w_2,w_3,w_4,w_5	Weyl group elements
11,138	$W(\tau)$	Whittaker function
5,6	$W_\nu(\tau,w)$	Whittaker function on GL(2)
42-43	$W^{(\nu_1,\nu_2)}_{n_1,n_2}(\tau,w)$	Whittaker function on GL(3) (if w omitted, $w = w_1$ understood)
17	x_1,x_2,x_3,x_4,y_1,y_2	Coordinates on $\mathcal{H}$
20	α,β,γ	
21,32,34	Δ_1,Δ_2	Differential operators
63,73	Γ	GL(3,$\mathbb{Z}$)
65	Γ_∞	Subgroup of Γ
65	Γ^2	Subgroup of Γ
65	Γ^2_1	Subgroup of Γ
65	Γ^2_∞	Subgroup of Γ
	Γ	Eulerian integral of second kind
18	ι	Involution of GL(3)
21	λ,μ	Eigenvalues of Δ_1,Δ_2
125	$\lambda(s)$	Rieman zeta function with Gamma factors
19	ν_1,ν_2	Complex parameters
63	ϕ	Automorphic form
71	$\tilde{\phi}$	Dual automorphic form
2,17	τ	Variable in $\mathcal{H}$
125	τ_ν	Divisor function
9,81	σ_ν	Divisor function
85-86	σ_{ν_1,ν_2}	Generalized divisor function
	ζ	Riemann zeta function

TABLE OF CONTENTS

CHAPTER I

AUTOMORPHIC FORMS ON GL(2)

In this chapter, we shall review those aspects of the theory of automorphic forms on GL(2) whose analogs on GL(3) will be studied in subsequent chapters. These topics include the theory of Whittaker functions, the special functions required for the Fourier expansions of automorphic forms; Fourier expansions of automorphic forms in general, and Eisenstein series in particular; the analytic continuation and functional equations of the Eisenstein series; Shintani's "explicit formula" for the p-adic Whittaker function, and the interpretation of Fourier coefficients as traces of finite-dimensional representations of the general linear group; Hecke operators; Euler products of L-series associated with automorphic forms; Mellin transforms of Whittaker functions; and the analytic continuation and functional equations of L-series associated with automorphic forms.

Although we shall now discuss these topics in the classical GL(2) case, our object is not so much to provide an introduction to what may already be a familiar topic, as to orient the reader as to our aims and methods in a more classical setting. Thus, in this chapter, many details and proofs may be elided. For further expository material on the subject matter of this chapter, we recommend Gelbart [4], and Terras [33].

Classical automorphic forms are functions on the upper half-plane. Two general types are known: holomorphic forms with weight, ommonly known as modular forms, and the real-analytic forms described Maass [22]. For many questions, such as the connection between the

Fourier expansions and the L-series, there is little difference between the Maass forms and holomorphic forms. The forms which we are to consider on GL(3) are precise analogs of the Maass forms on GL(2). It is with the latter that we shall be concerned in this chapter.

$G = GL(2,\mathbb{R})^{+}$ acts on $\mathcal{H} = \{x + iy \mid y>0\}$ as follows:

$$\begin{pmatrix} a & b \\ c & d \end{pmatrix} : \tau = x + iy \to \frac{a\tau + b}{c\tau + d}$$

The noneuclidean Laplacian is a G-invariant differential operator on $\mathcal{H}$:

$$\Delta = -y^2\left(\frac{d^2}{dx^2} + \frac{d^2}{dy^2}\right)$$

Let $\Gamma = SL(2,\mathbb{Z}) \subset G$. Let $\nu \in \mathbb{C}$. By an <u>automorphic form of type</u> ν we mean a function ϕ on $\mathcal{H}$ satisfying:

(1) $\phi(g.\tau) = \phi(\tau)$ for all $g \in \Gamma$;

(2) $\Delta\phi = \nu(1-\nu).\phi$;

(3) There exists a constant N such that $\phi(iy) = O(y^N)$ for y sufficiently large.

ϕ is called a <u>cusp form</u> if furthermore:

$$\int_0^1 \phi(x + iy)dx = 0 \quad \text{for all } y. \tag{1.1}$$

Let $\mathcal{M}(\nu)$ be the space of all automorphic forms of type ν, and let $\mathcal{M}_0(\nu)$ be the subspace of all cusp forms. Evidently, the spaces $\mathcal{M}(\nu)$ and $\mathcal{M}(1-\nu)$ are the same, because the condition (2) is symmetrical in ν and $1-\nu$.

We may also consider the Eisenstein series. If $\mathrm{re}(\nu) > 1$, let:

$$G_\nu(\tau) = \tfrac{1}{2}\pi^{-\nu}\Gamma(\nu)\Sigma \frac{y^\nu}{|m\tau+n|^{2\nu}}$$

(sum over $m, n \in \mathbb{Z}$, not both zero). Then G_ν is an automorphic form of type ν. It may be shown that $G_\nu(\tau)$ has meromorphic continuation to all values of ν, with simple poles at 0 and 1. Also, we have the functional equation:

$$G_\nu(\tau) = G_{1-\nu}(\tau). \tag{1.2}$$

We shall see shortly that this analytic continuation and functional equation for G_ν may be obtained through the Fourier expansions of the Eisenstein series. Taking this on faith for the moment, let $\mathcal{E}(\nu)$ be the one-dimensional space spanned by G_ν if $\nu \neq 0$ or 1, or by the constant functions on $\mathfrak{H}$ if $\nu = 0$ or 1. We may now state some basic facts about automorphic forms:

THEOREM. <u>We have</u> $\mathcal{M}(\nu) = \mathcal{M}_0(\nu) \oplus \mathcal{E}(\nu)$. <u>The space</u> $\mathcal{M}_0$ <u>is zero-dimensional except for some values of</u> ν <u>such that</u> $\mathrm{re}(\nu) = 1/2$. <u>In any case,</u> $\mathcal{M}(\nu)$ <u>is finite-dimensional</u>.

The values of ν for which cusp forms exist are scattered up and down the critical axis in a way which remains mysterious, although their asymptotic distribution may be studied using the trace formula. If we consider automorphic forms with respect to proper subgroups of finite index in Γ, it is conceivable that forms will also occur when ν is a real number between 0 and 1, although it is conjectured that this never occurs. No cases are known where the dimension of $\mathcal{M}_0$ is greater than one.

Let us now consider the Fourier expansions of automorphic forms. To this end, we introduce the topic of Whittaker functions.

Whittaker originally defined his functions as solutions of the confluent hypergeometric differential equation. Today, the term Whittaker function is used in a very general sense, including real and p-adic special functions, as well as functions defined on the adele group, for all reductive groups. If the group is GL(2), and the field is $\mathbb{R}$, the Whittaker functions will be those originally considered by Whittaker. Whittaker functions on GL(3) will be a major topic in this book. For more discussion of the Whittaker functions, see Jacquet [13], Schiffmann [29], Jacquet and Langlands [14], Shalika [28], Piatetski-Shapiro [24] and [25], Gelbart [4], Kostant [17], and Goodman and Wallach [6].

If ϕ is an automorphic form, then ϕ satisfies the three conditions:

(1') $\phi(\tau + 1) = \phi(\tau)$;

(2) $\Delta\phi = \nu(1 - \nu)\phi$;

(3) There exists a constant N such that $\phi(iy) = O(y^N)$ for y sufficiently large.

Let us construct special functions with these properties. Let n be an integer. First note that:

$$I_\nu^*(y) = \pi^{-\nu}\Gamma(\nu)y^\nu$$

is an eigenfunction for Δ (this is easy to see. The gamma factor is included for convenience). Let w be either of the two matrices:

$$w_0 = \begin{pmatrix} 1 & \\ & 1 \end{pmatrix} \qquad\qquad w_1 = \begin{pmatrix} & 1 \\ -1 & \end{pmatrix}$$

and let n be an integer. Intuitively, $W_n^\nu(\tau,w)$ will be that part of the function:

$$\tau \longrightarrow I_\nu^*(w\tau)$$

which transforms according to the rule:

$$(1'') \quad W_n^\nu(\tau+x) = e(-nx).W_n^\nu(\tau).$$

Here as always, we denote $e(x) = e^{2\pi ix}$.

If $w = w_0$, then $I_\nu^*(w.\tau) = I_\nu^*(\tau)$ is invariant with respect to translation by x, so we should define:

$$W_n^\nu(\tau, w_0) = \begin{cases} I_\nu^*(\tau) & \text{If } n = 0; \\ 0 & \text{Otherwise.} \end{cases}$$

On the other hand, if $w = w_1$, then we may separate out that part of $I^*(w.\tau)$ which transforms according to (1'') by integration. Thus, we define:

$$W_n^\nu(\tau, w_1) = \int_{-\infty}^{\infty} I^*(w_1.(\tau + x))e(-nx)dx$$

The integral being convergent if $re(\nu)>\frac{1}{2}$; however, this function has analytic continuation to all values of ν. It follows from the G-invariance of Δ that the Whittaker functions, as defined here, satisfy condition (2). If $n = 0$, we have:

$$W_0^\nu(\tau, w_1) = \pi^{\frac{1}{2}-\nu}\Gamma(\nu - \tfrac{1}{2})y^{1-\nu}. \tag{1.3}$$

On the other hand, if $n \neq 0$, we have:

$$W_n^\nu(\tau, w_1) = |n|^{\nu-1} W_1^\nu(n\tau, w_1) \tag{1.4}$$

where, if $n<0$ we interpret $W_1^\nu(n\tau, w_1)$ to mean $W_1^\nu(-n\bar{\tau}, w_1)$, a useful convention. Thus, in practice, we may only consider Whittaker functions where $n = 0$ or 1. If $n = 0$, the Whittaker function will be called <u>degenerate</u>. We see that there are two (nonzero) degenerate Whittaker functions, corresponding to $w = w_0$ and $w = w_1$, but only one nonzero nondegenerate Whittaker function, corresponding to $w = w_1$.

In standard notation, we have:

$$W_1^{\nu}(\tau, w_1) = 2\sqrt{y}\,K_{\nu-\frac{1}{2}}(2\pi y)e(x) = W_{0,\nu-\frac{1}{2}}(4\pi y)e(x). \tag{1.5}$$

Here $W_{0,\nu}(z)$ is Whittaker's first function, and $K_\nu(z)$ is a standard Bessel function. Cf. Watson [38], Whittaker and Watson [40], and Gradshteyn and Ryzhik [7], but note that Whittaker and Watson use a nonstandard notation for $K_\nu(z)$.

We have two functional equations (which require proof). $W_n^{\nu}(\tau,w)$ has meromorphic continuation to all values of ν, even analytic continuation if $w = w_0$, and:

$$\zeta(2\nu-1)W_0^{\nu}(\tau,w_1) = \zeta(2-2\nu)W_0^{1-\nu}(\tau,w_0) \tag{1.6}$$

$$W_1^{\nu}(\tau,w_1) = W_1^{1-\nu}(\tau,w_1) \tag{1.7}$$

Here ζ is Riemann's zeta function. The first functional equation could also be written using the Gamma function instead of the zeta function, using the functional equation of the Riemann zeta function.

Where no confusion is possible, we may write $W(\tau)$ for $W_1^{\nu}(\tau,w_1)$. It is technically important to have estimates for the magnitude of this function. Actually, an asymptotic series for this function is known (cf. Whittaker and Watson [40], Chapter XVI); however, we will obtain no analogous result on GL(3), so let us indicate another method of estimating $W(\tau)$, which is analogous to the procedure which we will follow on GL(3). This depends on knowing the Mellin transform:

$$\int_0^{\infty} W(iy)y^s\,\frac{dy}{y} = \tfrac{1}{2}\pi^{-\frac{1}{2}-s}\Gamma\left(\frac{s+\nu}{2}\right)\left(\frac{s+1-\nu}{2}\right) \tag{1.8}$$

Thus, by the Mellin inversion formula:

$$W(iy) = \frac{1}{2\sqrt{\pi}} \frac{1}{2\pi i} \int_{\sigma-i\infty}^{\sigma+i\infty} \Gamma(\frac{s+\nu}{2})\Gamma(\frac{s+1-\nu}{2})(\pi y)^{-s}ds \tag{1.9}$$

for σ sufficiently large. An integral of this type is known as a <u>Mellin-Barnes integral</u>.

The following estimate for the Gamma function follows from the result of Whittaker and Watson [40], 13.6:

$$|\Gamma(\sigma+it)| \sim \sqrt{2\pi}\, t^{\sigma-\frac{1}{2}} e^{\frac{-\pi t}{2}} \qquad \text{As} \quad t \to \infty \tag{1.10}$$

Thus, we may justify moving the line of integration either to the left (avoiding the poles of the Gamma function), or to the right. Moving to the left, and summing the residues in two infinite families, namely, when $s+\nu$ (resp. $s+1-\nu$) is a nonpositive even integer, we obtain:

$$\left.\begin{aligned} W(iy) = \frac{1}{2\sqrt{\pi}}\{\sum_{n=0}^{\infty} \frac{(-1)^n}{n!}\Gamma(\tfrac{1}{2}-\nu-n)(\pi y)^{2n+\nu} \\ +\sum_{n=0}^{\infty} \frac{(-1)^n}{n!}\,\Gamma(-\tfrac{1}{2}+\nu-n)(\pi y)^{2n+1-\nu}\}. \end{aligned}\right\} \tag{1.11}$$

This is equivalent to the expression of Whittaker's first function in terms of Whittaker's second function (cf. Whittaker and Watson, 16.41).

On the other hand, moving the line of integration to the right, we obtain the following estimate:

THEOREM: _If_ $n > 1 + re(\nu), 2 - re(\nu)$, _then_ $y^n.W(x+iy)$ _is bounded on the upper half-plane_.

Although it is possible to be far more precise, this estimate is sufficient for the convergence questions which arise in the study of automorphic forms. Differentiating under the integral sign, we have similar estimates for all derivatives of W.

Let us return to the Fourier expansions of the Eisenstein series:

$$G_\nu(\tau) = \zeta(2\nu)W_0^\nu(\tau, w_0) + \zeta(2\nu-1)W_0^\nu(\tau,w_1) + \tag{1.12}$$

$$\sum_{n\neq 0} |n|^{-\nu}\sigma_{2\nu-1}(|n|)W_1^\nu(n\tau,w_1),$$

here we denote:

$$\sigma_\nu(n) = \sum_{\substack{d|n \\ d>0}} d^\nu$$

The Fourier expansion (1.12) is to be understood in terms of the Bruhat decomposition of G. Let us explain. Given any matrix A in G, A may be written as $B_1.w.B_2$ where B_1, B_2 are upper triangular, and w = either w_0 or w_1. If c is the lower left-hand coefficient in the matrix A, then $w = w_1$ if $c \neq 0$, w_0 if $c = 0$. In the first case, A is said to be in the _Big Cell_ of the Bruhat decomposition; in the other case, A is said to be in the _Little Cell_. In general, the Bruhat decomposition is a cell decomposition of a given reductive group, parametrized by the elements of the Weyl group.

The Eisenstein series G_ν may be rewritten as:

$$G_\nu(\tau) = \pi^{-\nu}\Gamma(\nu)\zeta(2\nu) \sum_{g\in\Gamma_\infty\backslash\Gamma} \operatorname{im}(g.\tau)^\nu \qquad , \qquad (1.13)$$

where Γ_∞ is the group of all upper triangular unipotent matrices in Γ. We may split this into two parts, summing over both Bruhat cells. If this is done, we find that the contribution to the n-th Fourier coefficient of the terms in the big cell is:

$$\zeta(2\nu - 1)W_0^\nu(\tau,w_1) \qquad \text{if } n = 0;$$

$$|n|^{-\nu}\sigma_{2\nu-1}(|n|)W_1^\nu(n\tau,w_1) \qquad \text{otherwise.}$$

The contribution of the terms in the little cell is:

$$\zeta(2\nu)W_0^\nu(\tau,w_0) \qquad \text{if } n = 0$$

$$0 \qquad \text{otherwise.}$$

The actual evaluation of the Fourier terms may proceed from either the definition of G_ν or from (1.13). If (1.13) is the starting point, certain trigonometric sums arise, known as Ramanujan Sums (cf. Ramanujan [27]). With GL(3), as we shall see, the theory of these sums becomes quite interesting.

The analytic continuation and functional equations of the Eisenstein series follow from the Fourier expansion, together with the analytic continuation and functional equations (1.6-7) of the Whittaker functions.

Now, it develops that, in the nondegenerate case $n = 1$, the Whittaker function $W_1^{\nu}(\tau, w_1)$, which we shall henceforth abbreviate as simply $W(\tau)$, is uniquely characterized up to constant multiple by the conditions (1''), (2) and (3). This is known as the <u>local multiplicity one</u> theorem (in the real GL(2) case) -- a multiplicity one theorem is a uniqueness theorem for a suitably characterized Whittaker function. Definitive theorems of this type were obtained by Shalika [28], but see also Jacquet and Langlands [14], Gelbart [4], and Piatetski-Shapiro [24] and [25]. In the case at hand, the proof is rather trivial. On account of (1'') and (2), W satisfies a differential equation:

$$\left\{\frac{d^2}{dy^2} - \frac{1}{y^2}\,\nu(\nu-1) - 4\pi^2\right\} W(iy) = 0 \tag{1.14}$$

with asymptotically constant coefficients. This may be regarded as a perturbation of the equation:

$$\left\{\frac{d^2}{dy^2} - 4\pi^2\right\} W(iy) = 0$$

with solutions:

$$W(iy) = e^{\pm 2\pi y}.$$

Consequently, the Whittaker differential equation also has two solutions, one asymptotically large, and one asymptotically small. Condition (3) serves to force the latter solution.

This uniqueness assertion implies that <u>any</u> automorphic form of type ν has a Fourier expansion involving the Whittaker functions. Let us consider now the case of a cusp form ϕ. If $n \neq 0$, as the integral:

$$\int_0^1 \phi\left(\frac{iy}{n} + x\right)e^{-2\pi inx}dx$$

satisfies conditions (1''), (2) and (3), it is a constant multiple of $W(\tau)$. On the other hand, if $n = 0$, the corresponding Fourier term vanishes by (1.1). Thus, in this case too, we have a Fourier expansion:

$$\phi(\tau) = \sum_{n\neq 0} a_n W(n\tau)$$

Now, an important aspect of the theory of automorphic forms depends on the consideration of the L-series associated with ϕ. The Hecke operators form a commutative algebra of arithmetically defined operators on $\mathcal{M}_0(\nu)$, self-adjoined with respect to the Petersson inner product, whence $\mathcal{M}_0(\nu)$ has a basis of Hecke eigenforms. Assuming that ϕ is an eigenform, the Dirichlet series:

$$L(s,\phi) = \sum_{n=1}^{\infty} a_n n^{-s}$$

has an Euler product:

$$L(s,\phi) = \prod_p (1 - a_p p^{-s} + p^{-1-2s})^{-1}$$

The occurrence of the divisor sums in the Fourier expansion (1.12) of the Eisenstein series is well-known. What is less well-known is that the Fourier coefficients of the cusp forms may also be regarded as divisor sums! Let us explain why we make this assertion. The factors in the preceding Euler product are quadratic polynomials in p^{-s}. Let us factor these polynomials:

$$L(s,\phi) = \prod_p (1-a_p p^{-s} + p^{-1-2s})^{-1} =$$

$$\prod_p (1-\alpha_p p^{-s})^{-1}(1-\alpha'_p p^{-s})^{-1} =$$

$$\prod_p (1 + \alpha_p p^{-s} + \alpha_p^2 p^{-2s} + \dots)\ (1 + \alpha'_p p^{-s} + \alpha'^2_p p^{-2s} + \dots) =$$

$$\prod_p \sum_{k=0}^{\infty} \{ \sum_{k_1+k_2=k} \alpha_p^{k_1} \alpha_p'^{k_2} \} p^{-ks}$$

Expanding this infinite product, we obtain as the coefficient of n^{-s} a sum over the divisors of n. Thus, the coefficient a_n may be regarded as a divisor sum.

The point is that insight is always found by factoring the local factors in an Euler product into linear terms. Let us introduce the point of view that the Fourier coefficients should be regarded as special values of the characters of finite dimensional (algebraic) representations of $SL(2,\mathbb{C})$, a viewpoint which generalizes to $GL(n)$. When $n = 2$, this observation is rather trivial, because $SL(2,\mathbb{C})$ does not have many representations. Let V be the two-dimensional standard representation space of $SL(2,\mathbb{C})$. Then $SL(2,\mathbb{C})$ has precisely one representation of degree k for each positive integer k, namely, the symmetric power $\vee^{k-1}V$. Now, if $A \in SL(2,\mathbb{C})$ has eigenvalues α,α', then $\vee^{k-1}A$ has eigenvalues:

$$\alpha^k, \alpha^{k-1}\alpha', \ldots, \alpha'^k$$

Thus, if χ_k denotes the character of the representation $\vee^{k-1}V$, then:

$$\chi_k(A) = \sum_{k_1+k_2=k} \alpha^{k_1}\alpha'^{k_2} = \frac{\alpha^{k+1} - \alpha'^{k+1}}{\alpha - \alpha'} .$$

The second expression generalizes to a formula for the character of any algebraic representation of $SL(n,\mathbb{C})$ (cf. Weyl, [39], p. 201). This formula is the key to evaluating many Euler products which come up in the theory of automorphic forms, such as the Euler products which arise in the theory of the Rankin-Selberg method (cf. Jacquet, Piatetski-

Shapiro and Shalika, [16] and Friedberg [3]). The latter topic, however, is outside the scope of this book.

We see now that:

$$L(s,\phi) = \prod_p \sum_{k=0}^{\infty} \chi_k \begin{pmatrix} \alpha_p & \\ & \alpha_p' \end{pmatrix} \cdot p^{-ks}.$$

Thus, the Fourier coefficients may indeed be interpreted as character values.

One further important property of the L-series is their analytic continuations and functional equations. Let:

$$\Lambda(s,\phi) = \pi^{-\frac{1}{2}-s} \Gamma(\frac{s+\nu}{2})\Gamma(\frac{s+1-\nu}{2})L(s,\phi)$$

This is essentially the Mellin transform of ϕ . Owing to the invariance of ϕ with respect to the transformation:

$$\begin{pmatrix} & 1 \\ -1 & \end{pmatrix} : \tau \to -\frac{1}{\tau}$$

we find that $\Lambda(\phi,s)$ has analytic continuation to all values of s, and satisfies a functional equation:

$$\Lambda(\phi,s) = \Lambda(\phi,1-s)$$

We have given an indication of the range of topics which will concern us on GL(3). We shall devote much space to the properties of the GL(3) Whittaker functions, and much space to the Fourier coefficients of the Eisenstein series. In general, it may be stated that many questions become much richer, more complicated and interesting when one passes to GL(3). For example, the theory of Ramanujan sums, the relations between the coefficients arising from the Hecke algebra, the analytic continuations and functional equations of the Whittaker functions and the Eisenstein series -- these and many other topics only begin to show their full ramifications on GL(3). It may be argued, too, that even GL(3) is not fully typical of the situation to be encountered with higher rank groups.

CHAPTER II

THE DIFFERENTIAL EQUATIONS SATISFIED BY WHITTAKER FUNCTIONS

We introduce now some notations which will be standard throughout this entire book. Let $G = GL(3,\mathbb{R})$, let $K = O(3) \subset G$ be the maximal compact subgroup of orthogonal matrices, and let Z be the center of G, consisting of scalar matrices. We shall be concerned with functions on the homogeneous space $\mathcal{H} = G/ZK$, which plays the same role for $GL(3)$ as is played by the upper half plane for $GL(2)$. We shall always identify such a function on the homogeneous space with the corresponding function on G obtained by composition with the canonical map $G \longrightarrow G/ZK$.

Let us introduce coordinates on $\mathcal{H}$. The <u>Iwasawa decomposition</u> states that each coset in G/ZK has a unique representative of the form:

$$\tau = \begin{pmatrix} y_1y_2 & y_1x_2 & x_3 \\ & y_1 & x_1 \\ & & 1 \end{pmatrix}$$

where $y_1, y_2 > 0$. It is also useful to introduce an auxiliary coordinate x_4 defined by:

$$x_1x_2 = x_3 + x_4 \tag{2.1}$$

Let us now explain the reason for the auxiliary coordinate x_4. Let:

$$w_1 = \begin{pmatrix} & & -1 \\ & -1 & \\ -1 & & \end{pmatrix}$$

Then G possesses an involution:

$$\iota: g \to w_1 . {}^t g^{-1} . w_1$$

which preserves the Iwasawa decomposition, and hence induces an involution on $\mathcal{H}$, which we shall also denote as ι. In terms of the coordinates, ι has the effect:

$$\begin{array}{ll} x_1 \to -x_2 & x_3 \to x_4 \\ x_2 \to -x_1 & y_1 \to y_2 \\ x_3 \to x_4 & y_2 \to y_1 \end{array} \tag{2.2}$$

This is the reason for introducing the auxiliary coordinate x_4. The involution ι plays an important role in $GL(3)$ theory.

Let W be the group of the six matrices:

$$w_0 = \begin{pmatrix} 1 & & \\ & 1 & \\ & & 1 \end{pmatrix} \qquad w_3 = \begin{pmatrix} -1 & & \\ & & -1 \\ & -1 & \end{pmatrix}$$

$$w_1 = \begin{pmatrix} & & -1 \\ & -1 & \\ -1 & & \end{pmatrix} \qquad w_4 = \begin{pmatrix} & 1 & \\ & & 1 \\ 1 & & \end{pmatrix}$$

$$w_2 = \begin{pmatrix} & -1 & \\ -1 & & \\ & & -1 \end{pmatrix} \qquad w_5 = \begin{pmatrix} & & 1 \\ 1 & & \\ & 1 & \end{pmatrix}$$

We shall identify this group W with the Weyl Group of G.

We introduce an action of W on two complex variables ν_1, ν_2 as follows. This action is to be understood as generalizing the action $\nu \dashrightarrow 1-\nu$ of the $GL(2)$ Weyl group on one complex variable, which we saw in the functional equations of the $GL(2)$ Whittaker functions and Eisenstein series.

Let us consider the function:

$$I(\tau) = I_{(\nu_1,\nu_2)}(\tau) = y_1^{2\nu_1+\nu_2} y_2^{\nu_1+2\nu_2}$$

on $\mathcal{H}$, in terms of the coordinates. The reason for the peculiar exponents is that they make later formulae appear simpler. The action of W on the parameters ν_1, ν_2 is defined by requiring:

$$I_{(\nu_1-\frac{1}{3},\, \nu_2-\frac{1}{3})}(\tau) = I_{(\mu_1-\frac{1}{3},\, \mu_2-\frac{1}{3})}(w.\tau) \tag{2.3}$$

when:

$$(\mu_1, \mu_2) = w.(\nu_1, \nu_2) \tag{2.4}$$

if $x_1 = x_2 = x_3 = x_4 = 0$. Thus,

$$\begin{aligned}
w_0.(\nu_1, \nu_2) &= (\nu_1, \nu_2) \\
w_1.(\nu_1, \nu_2) &= (\tfrac{2}{3}-\nu_2,\ \tfrac{2}{3}-\nu_1) \\
w_2.(\nu_1, \nu_2) &= (\nu_1+\nu_2-\tfrac{1}{3}, \tfrac{2}{3}-\nu_2) \\
w_3.(\nu_1, \nu_2) &= (\tfrac{2}{3}-\nu_1, \nu_1+\nu_2-\tfrac{1}{3}) \\
w_4.(\nu_1, \nu_2) &= (1-\nu_1-\nu_2, \nu_1) \\
w_5.(\nu_1, \nu_2) &= (\nu_2, 1-\nu_1-\nu_2)
\end{aligned} \tag{2.5}$$

In addition to the two parameters ν_1, ν_2, it is convenient to introduce three equivalent parameters:

$$\begin{aligned}
\alpha &= -\nu_1-2\nu_2+1 \\
\beta &= -\nu_1+\nu_2 \\
\gamma &= 2\nu_1+\nu_2-1
\end{aligned} \tag{2.6}$$

we have:

$$\alpha+\beta+\gamma = 0. \tag{2.7}$$

Note that the action (2.4) of the Weyl group permutes these three quantities.

Let us now introduce the problem of the (nondegenerate) Whittaker functions. To maintain continuity of exposition, we shall first state the results of this chapter, then give the proofs.

We shall consider G-invariant differential operators on $\mathcal{H}$. These form a commutative algebra, isomorphic to a polynomial ring in two variables. Two particular generators Δ_1 and Δ_2 will be exhibited. Then:

$$\Delta_1 I = \lambda I \qquad \Delta_2 I = \mu I \tag{2.8}$$

where:

$$\lambda = -1-\beta\gamma-\gamma\alpha-\alpha\beta \qquad \mu = -\alpha\beta\gamma \tag{2.9}$$

We shall consider functions F on $\mathcal{H}$ which satisfy:

(1) F is an eigenfunction of Δ_1 and Δ_2 with the same eigenvalues as I, that is:

$$\Delta_1 F = \lambda F \qquad \Delta_2 F = \mu F \; ; \tag{2.10}$$

(2) We have:

$$F\left(\begin{pmatrix} 1 & x_2 & x_3 \\ & 1 & x_1 \\ & & 1 \end{pmatrix} \tau\right) = e(x_1+x_2).F(\tau) \qquad (2.11)$$

Because of Condition (2), the value of F at any $\tau \in \mathcal{H}$ is determined by the values when $x_1 = x_2 = x_3 = x_4 = 0$. Thus, we shall denote:

$$F(y_1,y_2) = F\left(\begin{pmatrix} y_1 y_2 & & \\ & y_1 & \\ & & 1 \end{pmatrix}\right) \qquad (2.12)$$

We shall see that the conditions (1) and (2) imply for F the differential equations:

$$\left\{y_1^2 \frac{\partial^2}{\partial y_1^2} + y_2^2 \frac{\partial}{\partial y_2^2} - y_1 y_2 \frac{\partial^2}{\partial y_1 \partial y_2} - 4\pi^2(y_1^2+y_2^2)\right\} F(y_1,y_2) = \lambda F(y_1,y_2); \qquad (2.13)$$

$$\left\{-y_1^2 y_2 \frac{\partial^3}{\partial y_1^2 \partial y_2} + y_1 y_2^2 \frac{\partial^3}{\partial y_1 \partial y_2^2} + 4\pi^2 y_1^2 y_2 \frac{\partial}{\partial y_2} - 4\pi^2 y_1 y_2^2 \frac{\partial}{\partial y_1} + y_1^2 \frac{\partial^2}{\partial y_1^2} - y_2^2 \frac{\partial^2}{\partial y_2^2} - 4\pi^2 y_1^2 + 4\pi^2 y_2^2\right\} F(y_1,y_2) = \mu F(y_1,y_2). \qquad (2.14)$$

Conversely, any solution to these differential equations corresponds in this fashion to a function satisfying (1) and (2).

(2.13-14) imply the further differential equation:

$$4\pi^2 y_1^2 y_2 \frac{\partial}{\partial y_2} F(y_1,y_2) = \left\{ y_1^3 \frac{\partial^3}{\partial y_1^3} - (4\pi^2 y_1^3 + \lambda y_1) \frac{\partial}{\partial y_1} + \mu + \lambda \right\} F(y_1,y_2) \tag{2.15}$$

Note that these equations are invariant under the action (2.4) of the Weyl group, since the parameters λ and μ are symmetric polynomials in α,β,γ, which are simply permuted.

Now, we shall see that the dimension of the space of solutions of the Whittaker differential equations (2.13-14) is at most six. Furthermore, except for some exceptional values of the parameters ν_1,ν_2, we may construct six linearly independent solutions of these differential equations, as generalized hypergeometric series. We shall use the notation:

$$\begin{aligned} a^{(n)} &= a(a+1)\ldots(a+n-1) = \frac{\Gamma(a+n)}{\Gamma(a)} \\ a_{(n)} &= a(a-1)\ldots(a-n+1) = \frac{\Gamma(a+1)}{\Gamma(a-n+1)} . \end{aligned} \tag{2.16}$$

Note that:

$$(-a)^{(n)} = (-1)^n a_{(n)} . \tag{2.17}$$

Define:

$$M_{(\nu_1,\nu_2)}(y_1,y_2) =$$

$$\sum_{n_1=0}^{\infty}\sum_{n_2=0}^{\infty} \frac{\left(\frac{3\nu_1+3\nu_2}{2}\right)^{(n_1+n_2)} (\pi y_1)^{2n_1}(\pi y_2)^{2n_2}}{n_1!n_2!\left(\frac{3\nu_1+1}{2}\right)^{(n_1)}\left(\frac{3\nu_2+1}{2}\right)^{(n_2)}\left(\frac{3\nu_1+3\nu_2}{2}\right)^{(n_1)}\left(\frac{3\nu_1+3\nu_2}{2}\right)^{(n_2)}} . \tag{2.18}$$

We shall see that $M_{(\nu_1,\nu_2)}$ is a solution of the differential equations (2.13-14). Thus, we obtain six solutions $M_{w.(\nu_1,\nu_2)}$, as w ranges through W, which will be linearly independent, unless

$$\nu_1 = \tfrac{1}{3},\ \nu_2 = \tfrac{1}{3} \text{ or } 1-\nu_1-\nu_2 = \tfrac{1}{3} . \tag{2.19}$$

Since we will show that the dimension of the space of functions satisfying (1) and (2) is at most six, it will follow that the dimension is exactly six, with the possible exceptions noted above. However, we do not expect that the dimension is properly less than six, even in the exceptional cases noted above.

It has been shown by Casselman and Zuckerman, and, independently by Kostant [17], that for more general groups G, the dimension of the space of solutions of the Whittaker differential equations is equal to the order of the Weyl group. This is the phenomenon just noted.

One particular other solution to (1) and (2) will be especially important for us. Let:

$$W(y_1,y_2) = W_{(\nu_1,\nu_2)}(y_1,y_2) = \tfrac{1}{4}\,\frac{1}{(2\pi i)^2}\int_{\sigma-i\infty}^{\sigma+i\infty}\int_{\sigma-i\infty}^{\sigma+i\infty} V(s_1,s_2)(\pi y_1)^{1-s_1} \cdot (\pi y_2)^{1-s_2}\,ds_1\,ds_2 \tag{2.20}$$

where the lines of integration are taken to the right of all poles of the integrand, and:

$$V(s_1,s_2) = \frac{\Gamma\left(\frac{s_1+\alpha}{2}\right)\Gamma\left(\frac{s_1+\beta}{2}\right)\Gamma\left(\frac{s_1+\gamma}{2}\right)\Gamma\left(\frac{s_2-\alpha}{2}\right)\Gamma\left(\frac{s_2-\beta}{2}\right)\Gamma\left(\frac{s_2-\gamma}{2}\right)}{\Gamma\left(\frac{s_1+s_2}{2}\right)} \tag{2.21}$$

The factor $\frac{1}{4}$ will eventually appear useful.

We shall show that W satisfies the differential equations (2.13-14). Moving the lines of integration to the left, and summing the residues at the poles of the integrand, one may express W as a linear combination of the six functions $M_{w.(\nu_1,\nu_2)}$. On the other hand, moving the lines of integration to the right, we have:

THEOREM 2.1. <u>There exist constants</u> N_1 <u>and</u> N_2, <u>depending in a continuous fashion on</u> ν_1,ν_2, <u>such that if</u> $n_1 > N_1$, $n_2 > N_2$, <u>then</u> $y_1^{n_1}.y_2^{n_2}.W(y_1,y_2)$ <u>is bounded on</u> $\mathcal{H}$.

Differentiating under the integral sign in (2.20) and moving the lines of integration to the right, one obtains similar estimates for all partial derivatives of W.

Let us recommend two exercises for the reader.

EXERCISE: If $a > 0$, $b < 0$, then $W(t^a,t^b)$ ----> 0 as t ----> 0 or ∞.

EXERCISE: Let $\Gamma = SL(2,\mathbb{Z})$, Γ_∞ be as in Chapter I. The sum:

$$\sum_{\left(\begin{smallmatrix} a & b \\ c & d \end{smallmatrix}\right) \in \Gamma_\infty\backslash\Gamma} W\left(\left(\begin{matrix} a & b & \\ c & d & \\ & & 1 \end{matrix}\right)\tau\right)$$

is convergent.

Of the six linearly independent solutions to the Whittaker differential equations, W is thus of rapid decay. We wish to use the fact that W is the only solution which satisfies a certain weak growth condition. This condition is:

(3) There exist constants n_1, n_2 such that:

$$y_1^{n_1} y_2^{n_2} W(y_1, y_2)$$

is bounded on the subset of $\mathcal{H}$ determined by the inequalities $y_1, y_2 > 1$.

THEOREM 2.2. *If* F *satisfies* (1), (2), (3), *then* F *is a constant multiple of* W.

This is a consequence of Shalika's local multiplicity one theorem for GL(n), in the real archimedean case (cf. Shalika [28], Gelbart [4] and Piatetski-Shapiro [24] and [25]). Unfortunately, we do not have a good proof of this fact, and we shall use Theorem 2.2 without proof.

If one wishes to see that this follows from Shalika's theorem, one argues as follows. The values (ν_1, ν_2) parametrize an induced representation of $GL(3,\mathbb{R})$ as follows: Let B be the group of upper triangular 3×3 nonsingular matrices. B has a unique character which is trivial on the center Z, and which has the value:

$$y_1^{2\nu_1+\nu_2-1} y_2^{\nu_1+2\nu_2-1}$$

on matrices of the form:

$$\begin{pmatrix} y_1y_2 & y_1x_2 & x_3 \\ & y_1 & x_1 \\ & & 1 \end{pmatrix}$$

Consider the representation of G induced by this character (cf. Godement-Jacquet [5]). By the local multiplicity one theorem, this representation has a unique Whittaker model corresponding to the character $e(x_1+x_2)$ of the group of unipotent matrices in B. The function F is assumed to lie in this Whittaker space. Furthermore, the Whittaker space has a unique one-dimensional subspace of K-stable vectors. Since W is right invariant with respect to K, it follows that W is uniquely characterized by conditions (1), (2) and (3).

We proceed now to the proofs of the preceding assertions. We shall require some elementary facts about Lie theory, for which we refer to Helgason [9], Varadarajan [36], and Lang [19].

Let $\mathfrak{g}$ be the Lie algebra of G, consisting of all $n \times n$ matrices, with Lie bracket operation $[X,Y] = XY-YX$ (matrix multiplication). $X \in \mathfrak{g}$ acts on the ring $C^\infty(G)$ of smooth functions on G via:

$$(Xf)(g) = \frac{d}{dt}f(g.\exp(tX))\Big|_{t=0} \tag{2.22}$$

X is a derivation in the sense that:

$$X(fg) = (Xf)g + f(Xg). \tag{2.23}$$

X may be regarded as a differential operator on G. The universal enveloping algebra $U(\mathfrak{g})$ of $\mathfrak{g}$ may be identified with the ring of operators on $C^\infty(G)$ generated by $X \in \mathfrak{g}$. Let . be the multiplication in $U(\mathfrak{g})$ (composition of operators). The Lie Bracket operation satisfies:

$$[X,Y] = X.Y-Y.X \ . \tag{2.24}$$

Similarly, we may consider $G_1 = SL(2,\mathbb{R})$, and $\mathfrak{g}_1$, the Lie algebra of G_1, consisting of matrices of trace zero, whose universal enveloping algebra may be regarded as a ring of differential operators on G_1 or, via the inclusion, on G.

Let us consider now the center $\mathfrak{Z}$ (resp. $\mathfrak{Z}_1$) of $U(\mathfrak{g})$ (resp. $U(\mathfrak{g}_1)$). Let D be a differential operator in $\mathfrak{Z}$. We claim that if f is a function on $\mathcal{H}$, that is, a function which is right invariant by K, then Df is also right invariant by K. Indeed, the right invariance of f with respect to K implies that f is invariant with respect to the Lie algebra of K acting as an algebra of differential operators. Since D commutes with the Lie algebra of K, Df is also invariant under the Lie algebra of K. Although K is not connected, this implies that at least Df is right invariant by the connected component of the identity of K. However, both connected components of K intersect the center Z of G, which also centralizes D, and consequently Df is right K-invariant.

We see, therefore, that $\mathfrak{Z}$ acts as an algebra of differential operators on $\mathcal{H}$. These operators may be seen to be invariant under the action of G on $\mathcal{H}$. It is known that the algebra $\mathfrak{Z}$ is commutative, and is a polynomial ring in three indeterminates, while $\mathfrak{Z}_1$ is a polynomial ring in two indeterminates. Let us consider now how to construct generators (compare Maass [22b]).

Let X_{ij} be the 3×3 matrix with a 1 at the i,j-th component, zeros elsewhere. Thus, the X_{ij} comprise a basis of $\mathfrak{g}$. Using Kronecker's delta, we have:

$$[X_{p,q}, X_{r,s}] = \delta_{q,r}X_{p,s} - \delta_{s,p}X_{r,q} \tag{2.25}$$

Now, if n is any positive integer, we claim that:

$$\sum_{i_1=1}^{3} \sum_{i_2=1}^{3} \cdots \sum_{i_n=1}^{3} X_{i_1 i_2} X_{i_2 i_3} \cdots X_{i_n i_1} \tag{2.26}$$

lies in the center of $U(\mathfrak{g})$. To prove this, we shall assume $n = 2$, to avoid confusing notations. Then:

$$[X_{pq}, \sum_{i_1} \sum_{i_2} X_{i_1 i_2} X_{i_2 i_1}] =$$

$$\sum_{i_1} \sum_{i_2} \{[X_{pq}, X_{i_1 i_2}] X_{i_2 i_1} + X_{i_1 i_2} [X_{pq}, X_{i_2 i_1}]\}$$

By (2.25), this equals:

$$\sum_{i_1}\sum_{i_2} (\delta_{qi_1}X_{pi_2}X_{i_2i_1} - \delta_{i_2p}X_{i_1q}X_{i_2i_1} + X_{i_1i_2}\delta_{qi_2}X_{pi_1} - X_{i_1i_2}\delta_{i_1p}X_{i_2q})$$

$$= \sum_{i_2} X_{pi_2}X_{i_2q} - \sum_{i_1} X_{i_1q}X_{pi_1} + \sum_{i_1} X_{i_1q}X_{pi_1} - \sum_{i_2} X_{pi_2}X_{i_2q} = 0 .$$

Thus (2.26) commutes with the generators X_{pq} of $U(\mathfrak{g})$, and lies in the center of $U(\mathfrak{g})$. The proof is similar if $n \neq 2$.

Now, let:

$$H_0 = \begin{pmatrix} \frac{1}{3} & & \\ & \frac{1}{3} & \\ & & \frac{1}{3} \end{pmatrix}$$

and let:

$$Y_{ij} = \begin{cases} X_{ii} - H_0 & \text{if } i = j ; \\ X_{ij} & \text{if } i \neq j . \end{cases} \tag{2.27}$$

Then, we see that, since H_0 lies in the center of $\mathfrak{g}$, hence of $U(\mathfrak{g})$,

$$\sum_{i_1=1}^{3} \cdots \sum_{i_n=1}^{3} Y_{i_1i_2}Y_{i_2i_3}\cdots Y_{i_ni_1} \tag{2.28}$$

is a sum of expressions such as (2.26) times powers of H_0. Consequently, (2.28) also lies in the center of $U(\mathfrak{g})$; furthermore, since Y_{ij} lies in $\mathfrak{g}_1$, (2.28) gives us elements of the center of $U(\mathfrak{g}_1)$. Now, we would like a more mnemonic basis for $\mathfrak{g}$ than the basis X_{ij}. Let:

$$H_0 = \begin{pmatrix} \frac{1}{3} & & \\ & \frac{1}{3} & \\ & & \frac{1}{3} \end{pmatrix} \qquad H_1 = \begin{pmatrix} \frac{1}{3} & & \\ & \frac{1}{3} & \\ & & -\frac{2}{3} \end{pmatrix} \qquad H_2 = \begin{pmatrix} \frac{2}{3} & & \\ & -\frac{1}{3} & \\ & & -\frac{1}{3} \end{pmatrix}$$

$$X_0 = \begin{pmatrix} 0 & 0 & 0 \\ 0 & 0 & 0 \\ 1 & 0 & 0 \end{pmatrix} \qquad X_1 = \begin{pmatrix} 0 & 0 & 0 \\ 0 & 0 & 1 \\ 0 & 0 & 0 \end{pmatrix} \qquad X_2 = \begin{pmatrix} 0 & 1 & 0 \\ 0 & 0 & 0 \\ 0 & 0 & 0 \end{pmatrix}$$

$$Z_0 = \begin{pmatrix} 0 & 0 & 1 \\ 0 & 0 & 0 \\ 0 & 0 & 0 \end{pmatrix} \qquad Z_1 = \begin{pmatrix} 0 & 0 & 0 \\ 0 & 0 & 0 \\ 0 & 1 & 0 \end{pmatrix} \qquad Z_2 = \begin{pmatrix} 0 & 0 & 0 \\ 1 & 0 & 0 \\ 0 & 0 & 0 \end{pmatrix}$$

Now, let us denote as $\mathcal{D}_1$, $\mathcal{D}_2$ the differential operators (2.28) when $n = 2$, $n = 3$ respectively. We wish to make these explicit. Noting that:

$$\begin{aligned} [X_0, Z_0] &= -H_1 - H_2 \qquad [H_1, H_2] = 0 \\ [X_1, Z_1] &= 2H_1 - H_2 \\ [X_2, Z_2] &= -H_1 + 2H_2 \end{aligned} \tag{2.29}$$

we find that:

$$\begin{aligned} \mathcal{D}_1 &= 2H_1^2 + 2H_2^2 - H_1H_2 - H_2H_1 + X_0Z_0 + Z_0X_0 + X_1Z_1 + Z_1X_1 + X_2Z_2 + Z_2X_2 \\ &= 2(H_1^2 - H_1 + H_2^2 - H_2 - H_1H_2 + Z_0X_0 + X_1Z_1 + X_2Z_2) \ . \end{aligned} \tag{2.30}$$

Now, let Δ_1 be the differential operator on $\mathcal{H} = G/ZK$ induced by the differential operator $\frac{1}{2}\mathfrak{D}_1$ on G. Note that if f is a function on $\mathcal{H}$, then, since f is right invariant by K, f is annihilated by the left ideal $U(\mathfrak{g}).\mathfrak{k}$ of $U(\mathfrak{g})$, where $\mathfrak{k}$ is the Lie algebra of K. A $\mathbb{Z}$-basis for $\mathfrak{k}$ consists of X_0-Z_0, X_1-Z_1, and X_2-Z_2. Thus, Δ_1 is equal, as a differential operator on $\mathcal{H}$ to:

$$H_1^2-H_1+H_2^2-H_2-H_1H_2+X_1^2+X_2^2+Z_0^2 \tag{2.31}$$

since this differs from $\frac{1}{2}\mathfrak{D}_1$ by an element of the **left ideal generated** by X_0-Z_0, X_1-Z_1, and X_2-Z_2.

Now, referring back to (2.22), we see that as differential operators on $\mathcal{H}$, the effects of the differential operators here are:

$$\begin{aligned} H_1 &= y_1\frac{\partial}{\partial y_1} \\ H_2 &= y_2\frac{\partial}{\partial y_2} \\ Z_0 &= y_1y_2\frac{\partial}{\partial x_3} \\ X_1 &= y_1\frac{\partial}{\partial x_1} + y_1x_2\frac{\partial}{\partial x_3} \\ X_2 &= y_2\frac{\partial}{\partial x_2} \end{aligned} \tag{2.32}$$

Consequently:

$$\Delta_1 = y_1^2 \frac{\partial^2}{\partial y_1^2} + y_2^2 \frac{\partial^2}{\partial y_2^2} - y_1 y_2 \frac{\partial^2}{\partial y_1 \partial y_2} + y_1^2(x_2^2+y_2^2) \frac{\partial^2}{\partial x_3^2}$$
$$+ y_1^2 \frac{\partial^2}{\partial x_1^2} + y_2^2 \frac{\partial^2}{\partial x_2^2} + 2y_1^2 x_2 \frac{\partial^2}{\partial x_1 \partial x_3} . \tag{2.33}$$

It follows that $I_{(\nu_1,\nu_2)}$ is an eigenfunction of Δ_1, with eigenvalue:

$$3(\nu_1^2+\nu_1\nu_2+\nu_2^2-\nu_1-\nu_2) = -1-\beta\gamma-\gamma\alpha-\alpha\beta = \lambda .$$

Now, if F satisfies (2), we have:

$$\frac{\partial F}{\partial x_1} = 2\pi i F$$
$$\frac{\partial F}{\partial x_2} = 2\pi i F \tag{2.34}$$
$$\frac{\partial F}{\partial x_3} = 0$$

from which we deduce (2.13).

Similarly, we find, using the commutation relations (2.29) (and some others not recorded there), that:

$$\mathfrak{D}_2 \equiv 3(-H_1^2H_2+H_1H_2^2+2H_1^2-H_1H_2-2H_1+Z_0X_2X_1+Z_0X_1X_2-X_1^2H_2+X_2^2H_1-Z_0^2H_1+Z_0^2H_2$$
$$+2X_1^2+Z_0^2) \tag{2.35}$$

modulo the left ideal $U(\mathfrak{g}).\mathfrak{k}$. Now, let:

$$\Delta_2 = \frac{1}{3}\mathcal{D}_2 - \Delta_1 \equiv -H_1^2H_2+H_1H_2^2+H_1^2-H_2^2-H_1+H_2-X_1^2H_2+X_2^2H_1-Z_0^2H_1+Z_0^2H_2+Z_0X_2X_1 + Z_0X_1X_2 \tag{2.36}$$

We find that:

$$\begin{aligned}\Delta_2 = & -y_1^2y_2 \frac{\partial^3}{\partial y_1^2 \partial y_2} + y_1y_2^2 \frac{\partial^3}{\partial y_1 \partial y_2^2} - y_1^3y_2^2 \frac{\partial^3}{\partial x_3^2 \partial y_1} + y_1y_2^2 \frac{\partial^3}{\partial x_2^2 \partial y_1} \\ & - 2y_1^2y_2x_2 \frac{\partial^3}{\partial x_1 \partial x_3 \partial y_2} + (-x_2^2+y_2^2)y_1^2y_2 \frac{\partial^3}{\partial x_3^2 \partial y_2} - y_1^2y_2 \frac{\partial^3}{\partial x_1^2 \partial y_2} \\ & + 2y_1^2y_2^2 \frac{\partial^3}{\partial x_1 \partial x_2 \partial x_3} + 2y_1^2y_2x_2 \frac{\partial^3}{\partial x_2 \partial x_3^2} \\ & + y_1^2 \frac{\partial^2}{\partial y_1^2} - y_2^2 \frac{\partial^2}{\partial y_2^2} + 2y_1^2x_2 \frac{\partial^2}{\partial x_1 \partial x_3} \\ & + (x_2^2+y_2^2)y_1^2 \frac{\partial^2}{\partial x_3^2} + y_1^2 \frac{\partial^2}{\partial x_1^2} - y_2^2 \frac{\partial^2}{\partial x_2^2}\end{aligned} \tag{2.37}$$

Consequently, $I_{(\nu_1,\nu_2)}$ is an eigenfunction of Δ_2, with eigenvalue:

$$-2\nu_1^3 - 3\nu_1^2\nu_2 + 3\nu_1\nu_2^2 + 2\nu_2^3 + 3\nu_1^2 - 3\nu_2^2 - \nu_1 + \nu_2 = -\alpha\beta\gamma = \mu .$$

Also, the proof of (2.14) now follows, similarly to the proof of (2.13).

Now, we observe that F satisfying:

$$(\Delta_1-\lambda)F = 0$$

$$(\Delta_2-\mu)F = 0$$

is also annihilated by:

$$(\Delta_2-\mu) + (1-y_1\frac{\partial}{\partial y_1})(\Delta_1-\lambda) .$$

This is (2.15). Using this relation, one may successively eliminate all occurrences of $\frac{\partial}{\partial y_2}$ from (2.13). It is easy to see that this procedure will yield a sixth degree differential equation in F involving only $\frac{\partial}{\partial y_1}$; the exact expression is hard to find because of the length of the computation. Let is be stated that:

$$\Big\{ y_1^6 \frac{\partial^6}{\partial y_1^6} + 3y_1^5 \frac{\partial^5}{\partial y_1^5} + [-\lambda y_1^4-12\pi^2 y_1^6] \frac{\partial^4}{\partial y_1^4} + [(4\lambda+2\mu)y_1^3 - 24\pi^2 y_1^5]\frac{\partial^3}{\partial y_1^3}$$
$$+ [(-12\lambda-6\mu+\lambda^2)y_1^2+12\pi^2\lambda y_1^4+48\pi^4 y_1^6] \frac{\partial^2}{\partial y_1^2} + [(24\lambda-2\lambda\mu+18\mu-3\lambda^2)y_1$$
$$-12\pi^2(\lambda+\mu)y_1^3+48\pi^2 y_1^5] \frac{\partial}{\partial y_1} + [-24\lambda-24\mu+4\lambda\mu+3\lambda^2+3\mu^2]$$
$$+12\pi^2(\lambda+\mu)y_1^2+16\pi^2(-\lambda-y_2^2)y_1^4-64\pi^2 y_1^6 \Big\} F = 0. \qquad (2.38)$$

We may now see that the dimension of the space of solutions of (1) and (2) is at most six dimensional for any (ν_1,ν_2). Indeed, this ordinary differential equation implies that the space of restrictions of solutions of (1) and (2) to the line $y_2 = c$, where c is constant, is at most six dimensional; and (2.15) shows that the evolution of such a solution in y_2 is determined by such a restriction. Consequently, (1) and (2) have a solution space which is at most six dimensional.

We turn now to the proof that (2.18) satisfies the differential equations (2.13) and (2.14). We denote:

$$a_{n_1,n_2} = \frac{\left(\frac{3\nu_1+3\nu_2}{2}\right)^{(n_1+n_2)}}{n_1!n_2!\left(\frac{3\nu_1+1}{2}\right)^{(n_1)}\left(\frac{3\nu_2+1}{2}\right)^{(n_2)}\left(\frac{3\nu_1+3\nu_2}{2}\right)^{(n_1)}\left(\frac{3\nu_1+3\nu_2}{2}\right)^{(n_2)}}$$

To satisfy (2.13), we must prove the recursion formula:

$$\{e_1(e_1-1) + e_2(e_2-1) - e_1e_2 - \lambda\}a_{n_1n_2} = 4(a_{n_1-1,n_2}+a_{n_1,n_2-1}) \quad (2.39)$$

where we denote:

$$e_1 = 2\nu_1 + \nu_2 + 2n_1$$

$$e_2 = \nu_1 + 2\nu_2 + 2n_2 .$$

The term in brackets in (2.39) equals $2A$, where:

$$A = 2n_1^2 + 2n_2^2 - 2n_1n_2 - n_1 - n_2 + 3n_1\nu_1 + 3n_2\nu_2$$

We have:

$$a_{n_1-1,n_2} = \left(\frac{3\nu_1+3\nu_2}{2}+n_1+n_2-1\right)^{-1}.n\left(\frac{3\nu_1+1}{2}+n_1-1\right)\left(\frac{3\nu_1+3\nu_2}{2}+n_1-1\right)a_{n_1,n_2}$$

with a similar expression for a_{n_1,n_2-1}. It follows that:

$$2(a_{n_1-1,n_2}+a_{n_1,n_2-1}) = (3\nu_1+3\nu_2+2n_1+2n_2-2)^{-1}\{n_1(3\nu_1+2n_1-1)$$

$$\cdot(3\nu_1+3\nu_2+2n_1-2) + n_2(3\nu_2+2n_2-1)$$

$$\cdot(3\nu_1+3\nu_2+2n_2-2)\} .$$

We find that $(3\nu_1+3\nu_2+2n_1+2n_2-2).A$ equals the expression in brackets in (2.40), so:

$$2(a_{n_1-1,n_2}+a_{n_1,n_2-1}) = A.a_{n_1,n_2}$$

as required.

The proof that (2.14) is satisfied is similar. The recursion requires verification of the identity:

$$[3\nu_1+3\nu_2+2n_1+2n_2-2][-e_1(e_1-1)e_2+e_1e_2(e_2-1)+e_1(e_1-1)-e_2(e_2-1)-\mu] =$$

$$2n_1(3\nu_1-1+2n_1)(3\nu_1+3\nu_2+2n_1-2)(-e_2+1) - 2n_2(3\nu_2-1+2n_2)(3\nu_1+3\nu_2+2n_2-2)$$

$$\cdot(-e_1+1) ,$$

which we leave to the reader.

We now turn to the proof that the Mellin-Barnes integral (2.20) satisfies the differential equations (2.13-14). This may be done in either of two different ways. By moving the lines of integration to the left, and summing the residues, one obtains (2.20) as a linear combination of the six functions $M_{w.(\nu_1,\nu_2)}$, as w ranges through the Weyl group. Since we have verified (2.13-14) for these functions, it follows that $W(\tau)$ also satisfies (2.13-14). However, it is pleasant to prove the differential equations directly, using Cauchy's theorem. We have:

$$\left\{y_1^2 \frac{\partial^2}{\partial y_1^2} + y_2^2 \frac{\partial^2}{\partial y_2^2} - y_1 y_2 \frac{\partial^2}{\partial y_1 \partial y_2} - 4\pi^2(y_1^2+y_2^2)\right\} \cdot$$

$$\frac{1}{4}\,\frac{1}{(2\pi i)^2} \int_{\sigma-i\infty}^{\sigma+i\infty} \int_{\sigma-i\infty}^{\sigma+i\infty} V(s_1,s_2)(\pi y_1)^{1-s_1}(\pi y_2)^{1-s_2}\, ds_1 ds_2 =$$

$$\frac{1}{4}\,\frac{1}{(2\pi i)^2} \int_{\sigma-i\infty}^{\sigma+i\infty} \int_{\sigma-i\infty}^{\sigma+i\infty} \Big\{ (-s_1+1)(-s_1)V(s_1,s_2)$$

$$+ (-s_2+1)(-s_2)V(s_1,s_2) - (-s_1+1)(-s_2+1)V(s_1,s_2)$$

$$- 4V(s_1+2,s_2) - 4V(s_1,s_2+2) \Big\} \cdot (\pi y_1)^{1-s_1}$$

$$\cdot (\pi y_2)^{1-s_2} ds_1 ds_2$$

We show that:

$$4V(s_1+2,s_2) + 4V(s_1,s_2+2) = (s_1^2 - s_1 s_2 + s_2^2 - \lambda - 1)V(s_1,s_2).$$

Indeed, since $\Gamma(s+1) = s.\Gamma(s)$, the left side equals:

$$4\left(\frac{s_1+s_2}{2}\right)^{-1}\left\{\left(\frac{s_1+\alpha}{2}\right)\left(\frac{s_1+\beta}{2}\right)\left(\frac{s_1+\gamma}{2}\right)+\left(\frac{s_2-\alpha}{2}\right)\left(\frac{s_2-\beta}{2}\right)\left(\frac{s_2-\gamma}{2}\right)\right\}V(s_1,s_2).$$

Since $\alpha + \beta + \gamma = 0$, this equals:

$$\frac{s_1^3+s_2^3+(\beta\gamma+\gamma\alpha+\alpha\beta)(s_1+s_2)}{s_1+s_2}V(s_1,s_2) = (s_1^2-s_1s_2+s_2^2-\lambda-1)V(s_1,s_2)\ .$$

Our previous expression now equals:

$$\frac{1}{(2\pi i)^2}\int_{\sigma-i\infty}^{\sigma+i\infty}\int_{\sigma-i\infty}^{\sigma+i\infty}\Big\{(-s_1+1)(-s_1)+(-s_2+1)(-s_2)$$

$$-(-s_1+1)(-s_2+1)-s_1^2+s_1s_2-s_2^2+\lambda-1\Big\}$$

$$\cdot V(s_1,s_2)(\pi y_1)^{-s_1+1}(\pi y_2)^{-s_2+1}ds_1ds_2$$

$$=\frac{\lambda}{(2\pi i)^2}\int_{\sigma-i\infty}^{\sigma+i\infty}\int_{\sigma-i\infty}^{\sigma+i\infty}V(s_1,s_2)(\pi y_1)^{-s_1+1}(\pi y_2)^{-s_2+1}ds_1ds_2 = \lambda W(y_1,y_2)\ ,$$

as required.

The proof of (2.14) is equally pleasant. We leave this as an exercise for the reader.

CHAPTER III

JACQUET'S WHITTAKER FUNCTIONS

The material in this chapter is due to Jacquet [13]. Jacquet introduced the Whittaker functions on an arbitrary Chevalley group as integrals generalizing the integrals $W_n^{\nu}(\tau,w)$ of Chapter I. Observing that the analytic continuation and functional equations of these Whittaker functions followed from the corresponding properties of the Eisenstein series, he set out to give proofs not depending on the Eisenstein series. We shall specialize his general results to the case at hand. Again, for expository reasons, we shall present the results of the chapter before the actual proofs.

Using notations introduced in the last chapter, by analogy with the approach taken in Chapter I to the construction of the GL(2) Whittaker functions, let us define the Whittaker functions on GL(3) so that, intuitively, $W_{n_1,n_2}^{(\nu_1,\nu_2)}(\tau,w)$, for $w \in W$, $\tau \in \mathcal{H}$, will be that part of:

$$\tau \to \pi^{-3\nu_1-3\nu_2+\frac{1}{2}}\Gamma\left(\frac{3\nu_1}{2}\right)\Gamma\left(\frac{3\nu_2}{2}\right)\Gamma\left(\frac{3\nu_1+3\nu_2-1}{2}\right)I_{(\nu_1,\nu_2)}(\tau) \tag{3.1}$$

which transforms according to the rule:

$$W\left(\begin{pmatrix}1 & x_2 & x_3\\ & 1 & x_1\\ & & 1\end{pmatrix}\tau\right) = e(n_1x_1+n_2x_2)W(\tau)\ , \tag{3.2}$$

the gamma factors being included for convenience. To this end, we need to calculate the effects of $w \in W$ on the coordinates $x_1, x_2, x_3, x_4, y_1, y_2$. That is to say, we should determine $x_1', x_2', x_3', x_4', y_1', y_2'$ related by (2.1) such that the matrices:

$$w\tau = w\begin{pmatrix} y_1y_2 & y_1x_2 & x_3 \\ & y_1 & x_1 \\ & & 1 \end{pmatrix}, \quad \tau' = \begin{pmatrix} y_1'y_2' & y_1'x_2' & x_3' \\ & y_1' & x_1' \\ & & 1 \end{pmatrix} \tag{3.3}$$

differ by an orthogonal similitude on the right. Actually, it is only necessary for our purposes to know y_1, y_2 -- cf. formulae **(3.47-52)** below. **From (3.47-57)**, we see that, denoting $I_{(\nu_1,\nu_2)}(w\tau) = I_{(\nu_1,\nu_2)}(\tau')$:

$$I_{(\nu_1,\nu_2)}(w_0\tau) = I_{(\nu_1,\nu_2)}(\tau); \tag{3.4}$$

$$I_{(\nu_1,\nu_2)}(w_1\tau) = (x_3^2+x_2^2y_1^2+y_1^2y_2^2)^{-\frac{3\nu_1}{2}}(x_4^2+x_1^2y_2^2+y_1^2y_2^2)^{-\frac{3\nu_2}{2}}I_{(\nu_1,\nu_2)}(\tau); \tag{3.5}$$

$$I_{(\nu_1,\nu_2)}(w_2\tau) = (x_2^2+y_2^2)^{-\frac{3\nu_2}{2}}I_{(\nu_1,\nu_2)}(\tau); \tag{3.6}$$

$$I_{(\nu_1,\nu_2)}(w_3\tau) = (x_1^2+y_1^2)^{-\frac{3\nu_1}{2}}I_{(\nu_1,\nu_2)}(\tau); \tag{3.7}$$

$$I_{(\nu_1,\nu_2)}(w_4\tau) = (x_3^2+x_2^2y_1^2+y_1^2y_2^2)^{-\frac{3\nu_1}{2}}(x_2^2+y_2^2)^{-\frac{3\nu_2}{2}}I_{(\nu_1,\nu_2)}(\tau); \tag{3.8}$$

$$I_{(\nu_1,\nu_2)}(w_5\tau) = (x_1^2+y_1^2)^{-\frac{3\nu_1}{2}}(x_4^2+x_1^2y_2^2+y_1^2y_2^2)^{-\frac{3\nu_2}{2}} I_{(\nu_1,\nu_2)}(\tau) \ . \qquad (3.9)$$

Thus, we should define, if $re(\nu_1),\ re(\nu_2) > \frac{1}{3}$:

$$W_{n_1,n_2}^{(\nu_1,\nu_2)}(\tau,w_0) = \begin{cases} \pi^{-3\nu_1-3\nu_2+\frac{1}{2}} \cdot \Gamma(\frac{3\nu_1}{2})\Gamma(\frac{3\nu_2}{2})\Gamma(\frac{3\nu_1+3\nu_2-1}{2}) \cdot I_{(\nu_1,\nu_2)}(\tau) & \text{if } n_1 = n_2 = 0; \\ 0 & \text{otherwise.} \end{cases} \qquad (3.10)$$

$$W_{n_1,n_2}^{(\nu_1,\nu_2)}(\tau,w_1) =$$

$$\pi^{-3\nu_1-3\nu_2+\frac{1}{2}} \cdot \Gamma(\frac{3\nu_1}{2})\Gamma(\frac{3\nu_2}{2})\Gamma(\frac{3\nu_1+3\nu_2-1}{2}) \cdot I_{(\nu_1,\nu_2)}(\tau)\ e(n_1x_1+n_2x_2)$$

$$\int_{-\infty}^{\infty}\int_{-\infty}^{\infty}\int_{-\infty}^{\infty} (\xi_3^2+\xi_2^2y_1^2+y_1^2y_2^2)^{-3\nu_1/2} \cdot (\xi_4^2+\xi_1^2y_2^2+y_1^2y_2^2)^{-3\nu_2/2}$$

$$\cdot e(-n_1\xi_1-n_2\xi_2)d\xi_1 d\xi_2 d\xi_3 \ . \qquad (3.11)$$

$$W_{n_1,n_2}^{(\nu_1,\nu_2)}(\tau,w_2) =$$

$$\begin{cases} \pi^{-3\nu_1-3\nu_2+\frac{1}{2}} \cdot \Gamma(\frac{3\nu_1}{2})\Gamma(\frac{3\nu_2}{2})\Gamma(\frac{3\nu_1+3\nu_2-1}{2}) \cdot I_{(\nu_1,\nu_2)}(\tau)\ e(n_2x_2) \ . \\ \int_{-\infty}^{\infty} (\xi_2^2+y_2^2)^{-3\nu_2/2} \cdot e(-n_2\xi_2)d\xi_2 & \text{if } n_1 = 0; \\ 0 & \text{otherwise.} \end{cases} \qquad (3.12)$$

$$W_{n_1,n_2}^{(\nu_1,\nu_2)}(\tau,w_3) =$$

$$\pi^{-3\nu_1-3\nu_2+\frac{1}{2}}\cdot\Gamma(\frac{3\nu_1}{2})\Gamma(\frac{3\nu_2}{2})\Gamma(\frac{3\nu_1+3\nu_2-1}{2})\cdot I_{(\nu_1,\nu_2)}(\tau)\cdot e(n_1x_1)\ .$$

$$\int_{-\infty}^{\infty}(\xi_1^2+y_1^2)^{-3\nu_1/2}\cdot e(-n_1\xi_1)d\xi_1 \qquad \text{if} \quad n_2 = 0;$$

$$0 \qquad \text{otherwise.} \qquad (3.13)$$

$$W_{n_1,n_2}^{(\nu_1,\nu_2)}(\tau,w_4) =$$

$$\pi^{-3\nu_1-3\nu_2+\frac{1}{2}}\cdot\Gamma(\frac{3\nu_1}{2})\Gamma(\frac{3\nu_2}{2})\Gamma(\frac{3\nu_1+3\nu_2-1}{2})\cdot I_{(\nu_1,\nu_2)}(\tau)\cdot e(n_2x_2)\ .$$

$$\int_{-\infty}^{\infty}\int_{-\infty}^{\infty}(\xi_3^2+\xi_2^2y_1^2+y_1^2y_2^2)^{-3\nu_1/2}\cdot(\xi_2^2+y_2^2)^{-3\nu_2/2}e(-n_2\xi_2)d\xi_2d\xi_3$$

$$\text{if} \quad n_1 = 0;$$

$$0 \qquad \text{otherwise.} \qquad (3.14)$$

$$W_{n_1,n_2}^{(\nu_1,\nu_2)}(\tau,w_5) =$$

$$\pi^{-3\nu_1-3\nu_2+\frac{1}{2}}\cdot\Gamma(\frac{3\nu_1}{2})\Gamma(\frac{3\nu_2}{2})\Gamma(\frac{3\nu_1+3\nu_2-1}{2})\cdot I_{(\nu_1,\nu_2)}(\tau)\cdot e(n_1x_1)$$

$$\int_{-\infty}^{\infty}\int_{-\infty}^{\infty}(\xi_1^2+y_1^2)^{-3\nu_1/2}\cdot(\xi_4^2+\xi_1^2y_2^2+y_1^2y_2^2)^{-3\nu_2/2}e(-n_1\xi_1)d\xi_1d\xi_4$$

$$\text{if} \quad n_2 = 0;$$

$$0 \qquad \text{otherwise.} \qquad (3.15)$$

The absolute convergence of these integrals when $re(\nu_1)$, $re(\nu_2) > \frac{1}{3}$ follows from iterated applications of the integral formula (3.54) below.

Since the differential operators Δ_1 and Δ_2 of Chapter II are invariant under the action of $GL(3,\mathbb{R})$, and since these Whittaker functions are built up from translates of the Eigenfunction $I_{(\nu_1,\nu_2)}$, it follows that the differential equations (2.10) are satisfied by these Whittaker functions.

Now, let us note the following formulae, which, as in the GL(2) case, may be proved using a simple change of variables:

$$W^{(\nu_1,\nu_2)}_{n_1,n_2}(\tau,w_1) = |n_1|^{\nu_1+2\nu_2-2}|n_2|^{2\nu_1+\nu_2-2}\cdot W^{(\nu_1,\nu_2)}_{1,1}\left(\begin{pmatrix} n_1n_2 & & \\ & n_1 & \\ & & 1\end{pmatrix}\tau,w_1\right) \tag{3.16}$$

$$W^{(\nu_1,\nu_2)}_{n_1,0}(\tau,w_1) = |n_1|^{\nu_1+2\nu_2-2}\cdot W^{(\nu_1,\nu_2)}_{1,0}\left(\begin{pmatrix} n_1 & & \\ & n_1 & \\ & & 1\end{pmatrix}\tau,w_1\right) \tag{3.17}$$

$$W^{(\nu_1,\nu_2)}_{0,n_2}(\tau,w_1) = |n_2|^{2\nu_1+\nu_2-2}\cdot W^{(\nu_1,\nu_2)}_{0,1}\left(\begin{pmatrix} n_2 & & \\ & 1 & \\ & & 1\end{pmatrix}\tau,w_1\right) \tag{3.18}$$

$$W^{(\nu_1,\nu_2)}_{0,n_2}(\tau,w_2) = |n_2|^{-\nu_1+\nu_2-1}\cdot W^{(\nu_1,\nu_2)}_{0,1}\left(\begin{pmatrix} n_2 & & \\ & 1 & \\ & & 1\end{pmatrix}\tau,w_2\right) \tag{3.19}$$

$$W^{(\nu_1,\nu_2)}_{n_1,0}(\tau,w_3) = |n_1|^{\nu_1-\nu_2-1}\cdot W^{(\nu_1,\nu_2)}_{1,0}\left(\begin{pmatrix} n_1 & & \\ & n_1 & \\ & & 1\end{pmatrix}\tau,w_3\right) \tag{3.20}$$

$$W^{(\nu_1,\nu_2)}_{0,n_2}(\tau,w_4) = |n_2|^{2\nu_1+\nu_2-2}\cdot W^{(\nu_1,\nu_2)}_{0,1}\left(\begin{pmatrix} n_2 & & \\ & 1 & \\ & & 1\end{pmatrix}\tau,w_4\right) \tag{3.21}$$

$$W^{(\nu_1,\nu_2)}_{n_1,\,0}(\tau,w_5) = |n_1|^{\nu_1+2\nu_2-2}\cdot W^{(\nu_1,\nu_2)}_{1\,,\,0}\left(\begin{pmatrix} n_1 & & \\ & n_1 & \\ & & 1\end{pmatrix}\tau, w_5\right) \qquad (3.22)$$

Thus, we are reduced to thirteen basic Whittaker functions, namely, those nonvanishing ones where n_1, n_2 are each either 0 or 1.

The function $W^{(\nu_1,\nu_2)}_{1\,,\,1}(\tau,w_1)$ will be called nondegenerate; the other twelve (with n_1 or $n_2 = 0$) will be called degenerate.

Let us also record the following relations involving the involution ι:

$$W^{(\nu_1,\nu_2)}_{0\,,\,0}(\tau,w_0) = W^{(\nu_2,\nu_1)}_{0\,,\,0}({}^{\iota}\tau,w_0) \qquad (3.23)$$

$$W^{(\nu_1,\nu_2)}_{n_1,n_2}(\tau,w_1) = W^{(\nu_2,\nu_1)}_{-n_2,-n_1}({}^{\iota}\tau,w_1) \qquad (3.24)$$

$$W^{(\nu_1,\nu_2)}_{0\,,n_2}(\tau,w_2) = W^{(\nu_2,\nu_1)}_{-n_2,\,0}({}^{\iota}\tau,w_3) \qquad (3.25)$$

$$W^{(\nu_1,\nu_2)}_{n_1,\,0}(\tau,w_3) = W^{(\nu_2,\nu_1)}_{0\,,-n_1}({}^{\iota}\tau,w_2) \qquad (3.26)$$

$$W^{(\nu_1,\nu_2)}_{0\,,n_2}(\tau,w_4) = W^{(\nu_2,\nu_1)}_{-n_2,\,0}({}^{\iota}\tau,w_5) \qquad (3.27)$$

$$W^{(\nu_1,\nu_2)}_{n_1,\,0}(\tau,w_5) = W^{(\nu_2,\nu_1)}_{0\,,-n_1}({}^{\iota}\tau,w_4) \qquad (3.28)$$

The main results of this chapter will be the analytic continuations and functional equations of the Whittaker functions. To be precise, each $W^{(\nu_1,\nu_2)}_{n_1,n_2}(\tau,w)$, originally defined for $re(\nu_1)$, $re(\nu_2) > \frac{1}{3}$, has meromorphic continuation to all ν_1,ν_2, indeed analytic continuation in the case of the nondegenerate function $W^{(\nu_1,\nu_2)}_{1,1}(\tau,w_1)$. As in the GL(2) case, the functional equations of the degenerate Whittaker functions involve a factor, which may be expressed either in terms of the Riemann zeta function, or the gamma function. We will find the former expressions more to our convenience.

$W^{(\nu_1,\nu_2)}_{1,1}(\tau,w_1)$ has analytic continuation to all values of (ν_1,ν_2), and for each $w \in W$, if $w\cdot(\nu_1,\nu_2) = (\mu_1,\mu_2)$, we have:

$$W^{(\nu_1,\nu_2)}_{1,1}(\tau,w_1) = W^{(\mu_1,\mu_2)}_{1,1}(\tau,w_1) \tag{3.29}$$

Thus, the nondegenerate Whittaker functions are entire functions of (ν_1,ν_2), and are invariant under the action (2.4). By contrast, the degenerate functions are only meromorphic. They are permuted by the action (2.4).

We have the following functional equations, from which others may be deduced (recall that w_2 and w_3 generate the Weyl group):

If $(\mu_1,\mu_2) = w_2\cdot(\nu_1,\nu_2)$ then:

$$\zeta(3\mu_2-1)\cdot W^{(\mu_1,\mu_2)}_{0,0}(\tau,w_5) = \zeta(3\nu_2)\cdot W^{(\nu_1,\nu_2)}_{0,0}(\tau,w_3) \tag{3.30}$$

If $(\mu_1,\mu_2) = w_3\cdot(\nu_1,\nu_2)$ then:

$$\zeta(3\mu_1-1)\cdot W^{(\mu_1,\mu_2)}_{0,0}(\tau,w_1) = \zeta(3\nu_1)\cdot W^{(\nu_1,\nu_2)}_{0,0}(\tau,w_5) \tag{3.31}$$

If $(\mu_1,\mu_2) = w_2\cdot(\nu_1,\nu_2)$ then:

$$\zeta(3\mu_2)\cdot W_{0,0}^{(\mu_1,\mu_2)}(\tau,w_4) = \zeta(3\nu_2-1)\cdot W_{0,0}^{(\nu_1,\nu_2)}(\tau,w_1) \tag{3.32}$$

If $(\mu_1,\mu_2) = w_3\cdot(\nu_1,\nu_2)$ then:

$$\zeta(3\mu_1)\cdot W_{0,0}^{(\mu_1,\mu_2)}(\tau,w_2) = \zeta(3\nu_1-1)\cdot W_{0,0}^{(\nu_1,\nu_2)}(\tau,w_4) \tag{3.33}$$

If $(\mu_1,\mu_2) = w_2\cdot(\nu_1,\nu_2)$ then:

$$\zeta(3\mu_2)\cdot W_{0,0}^{(\mu_1,\mu_2)}(\tau,w_0) = \zeta(3\nu_2-1)\cdot W_{0,0}^{(\nu_1,\nu_2)}(\tau,w_2) \tag{3.34}$$

If $(\mu_1,\mu_2) = w_3\cdot(\nu_1,\nu_2)$ then:

$$\zeta(3\mu_1-1)\cdot W_{0,0}^{(\mu_1,\mu_2)}(\tau,w_3) = \zeta(3\nu_1)\cdot W_{0,0}^{(\nu_1,\nu_2)}(\tau,w_0) \tag{3.35}$$

Thus W, acting on (ν_1,ν_2), permutes transitively the six functions:

$$\zeta(3\nu_1)\zeta(3\nu_2)\zeta(3\nu_1+3\nu_2-1)\cdot W_{0,0}^{(\nu_1,\nu_2)}(\tau,w_0)$$

$$\zeta(3\nu_1)\zeta(3\nu_2-1)\zeta(3\nu_1+3\nu_2-1)\cdot W_{0,0}^{(\nu_1,\nu_2)}(\tau,w_2)$$

$$\zeta(3\nu_1-1)\zeta(3\nu_2)\zeta(3\nu_1+3\nu_2-2)\cdot W_{0,0}^{(\nu_1,\nu_2)}(\tau,w_4)$$

$$\zeta(3\nu_1-1)\zeta(3\nu_2-1)\zeta(3\nu_1+3\nu_2-2)\cdot W^{(\nu_1,\nu_2)}_{0,0}(\tau,w_1)$$

$$\zeta(3\nu_1)\zeta(3\nu_2-1)\zeta(3\nu_1+3\nu_2-2)\cdot W^{(\nu_1,\nu_2)}_{0,0}(\tau,w_5)$$

$$\zeta(3\nu_1-1)\zeta(3\nu_2)\zeta(3\nu_1+3\nu_2-1)\cdot W^{(\nu_1,\nu_2)}_{0,0}(\tau,w_3)$$

We have a similar situation with the "partially degenerate" functions $W^{(\nu_1,\nu_2)}_{1,0}$ and $W^{(\nu_1,\nu_2)}_{0,1}$:

If $(\mu_1,\mu_2) = w_2\cdot(\nu_1,\nu_2)$ then: (3.36)

$$W^{(\mu_1,\mu_2)}_{1,0}(\tau,w_1) = W^{(\nu_1,\nu_2)}_{1,0}(\tau,w_1)$$

If $(\mu_1,\mu_2) = w_3\cdot(\nu_1,\nu_2)$ then: (3.37)

$$\zeta(3\mu_1-1)\cdot W^{(\mu_1,\mu_2)}_{1,0}(\tau,w_1) = \zeta(3\nu_1)\cdot W^{(\nu_1,\nu_2)}_{1,0}(\tau,w_5)$$

If $(\mu_1,\mu_2) = w_2\cdot(\nu_1,\nu_2)$ then: (3.38)

$$\zeta(3\mu_2-1)\cdot W^{(\mu_1,\mu_2)}_{1,0}(\tau,w_5) = \zeta(3\nu_2)\cdot W^{(\nu_1,\nu_2)}_{1,0}(\tau,w_3)$$

If $(\mu_1,\mu_2) = w_3\cdot(\nu_1,\nu_2)$ then: (3.39)

$$W^{(\mu_1,\mu_2)}_{1,0}(\tau,w_3) = W^{(\nu_1,\nu_2)}_{1,0}(\tau,w_3)$$

Thus W permutes transitively the three functions:

$$\zeta(3\nu_2)\zeta(3\nu_1+3\nu_2-1)\cdot W_{1,0}^{(\nu_1,\nu_2)}(\tau,w_3)$$

$$\zeta(3\nu_1)\zeta(3\nu_2-1)\cdot W_{1,0}^{(\nu_1,\nu_2)}(\tau,w_5)$$

$$\zeta(3\nu_1-1)\zeta(3\nu_1+3\nu_2-2)\cdot W_{1,0}^{(\nu_1,\nu_2)}(\tau,w_1)$$

In terms of the Bessel function K_ν, (cf. Watson [38]), we have explicitly:

$$W_{1,0}^{(\nu_1,\nu_2)}(\tau,w_3) = 2\pi^{-\frac{3\nu_2}{2} - 3\nu_2+\frac{1}{2}}\cdot\Gamma(\frac{3\nu_2}{2})\Gamma(\frac{3\nu_1+3\nu_2-1}{2})$$

$$\cdot y_1^{\frac{1}{2}+\frac{\nu_1}{2}+\nu_2}\cdot y_2^{\nu_1+2\nu_2}\cdot e(x_1)\cdot K_{\frac{3\nu_1-1}{2}}(2\pi y_1) \tag{3.40}$$

If $(\mu_1,\mu_2) = w_3\cdot(\nu_1,\nu_2)$ then: (3.41)

$$W_{0,1}^{(\mu_1,\mu_2)}(\tau,w_1) = W_{0,1}^{(\nu_1,\nu_2)}(\tau,w_1)$$

If $(\mu_1,\mu_2) = w_2\cdot(\nu_1,\nu_2)$ then: (3.42)

$$\zeta(3\mu_2-1)\cdot W_{0,1}^{(\mu_1,\mu_2)}(\tau,w_1) = \zeta(3\nu_2)\cdot W_{0,1}^{(\nu_1,\nu_2)}(\tau,w_4)$$

If $(\mu_1,\mu_2) = w_3\cdot(\nu_1,\nu_2)$ then: (3.43)

$$\zeta(3\mu_1-1)\cdot W^{(\mu_1,\mu_2)}_{0,1}(\tau,w_4) = \zeta(3\nu_1)\cdot W^{(\nu_1,\nu_2)}_{0,1}(\tau,w_2)$$

If $(\mu_1,\mu_2) = w_2\cdot(\nu_1,\nu_2)$ then: (3.44)

$$W^{(\mu_1,\mu_2)}_{0,1}(\tau,w_2) = W^{(\nu_1,\nu_2)}_{0,1}(\tau,w_2)$$

Thus W permutes transitively the three functions:

$$\zeta(3\nu_1)\zeta(3\nu_1+3\nu_2-1)\cdot W^{(\nu_1,\nu_2)}_{0,1}(\tau,w_2)$$

$$\zeta(3\nu_1-1)\zeta(3\nu_2)\cdot W^{(\nu_1,\nu_2)}_{0,1}(\tau,w_4)$$

$$\zeta(3\nu_2-1)\zeta(3\nu_1+3\nu_2-2)\cdot W^{(\nu_1,\nu_2)}_{0,1}(\tau,w_1)$$

We have the following explicit formula:

$$W^{(\nu_1,\nu_2)}_{0,1}(\tau,w_2) = 2\pi^{-3\nu_1-\frac{3\nu_2}{2}+\frac{1}{2}}\cdot\Gamma\left(\frac{3\nu_1}{2}\right)\Gamma\left(\frac{3\nu_1+3\nu_2-1}{2}\right)$$

$$\cdot y_1^{2\nu_1+\nu_2}\cdot y_2^{\frac{1}{2}+\nu_1+\frac{\nu_2}{2}}\cdot e(x_2)\cdot K_{\frac{3\nu_2-1}{2}}(2\pi y_2) \qquad (3.45)$$

Thus, in view of the functional equations, the degenerate Whittaker functions are given explicitly from (3.9), (3.39) and (3.44). On the other hand, we shall prove that there exists a constant $c(\nu_1,\nu_2)$ such that:

$$W_{1,1}^{(\nu_1,\nu_2)}(\tau,\ w_1) = c(\nu_1,\nu_2)\,W(y_1,y_2)e(x_1+x_2) \tag{3.46}$$

where $W(y_1,y_2)$ is defined by (2.20). This result will be superseded by the result of Chapter X, where we shall show that actually $c(\nu_1,\nu_2) = 1$ independent of ν_1,ν_2.

Another expression for the nondegenerate Whittaker function, as an integral of a Bessel function, may be found in Vinogradov and Takhtadzhyan [37].

We turn now to the proofs. Let us first consider (3.3-8). These follow from the following formulae. With x_1',x_2',x_3',y_1',y_2' as in (3.3), we will show that:

$$\text{If}\quad w = w_0,\ y_1' = y_1,\ y_2' = y_2; \tag{3.47}$$

$$\text{If}\quad w = w_1,\ y_1' = y_1\frac{(x_4^2+x_1^2y_2^2+y_1^2y_2^2)^{\frac{1}{2}}}{x_3^2+x_2^2y_1^2+y_1^2y_2^2},\ y_2' = y_2\frac{(x_3^2+x_2^2y_1^2+y_1^2y_2^2)^{\frac{1}{2}}}{x_4^2+x_1^2y_2^2+y_1^2y_2^2}; \tag{3.48}$$

$$\text{If}\quad w = w_2,\ y_1' = y_1\,(x_2^2+y_2^2)^{\frac{1}{2}},\ y_2' = y_2\,(x_2^2+y_2^2)^{-1}; \tag{3.49}$$

$$\text{If}\quad w = w_3,\ y_1' = y_1\,(x_1^2+y_1^2)^{-1},\ y_2' = y_2\,(x_1^2+y_1^2)^{\frac{1}{2}}; \tag{3.50}$$

$$\text{If } w = w_4,\ y_1' = y_1 \frac{(x_2^2+y_2^2)^{\frac{1}{2}}}{x_3^2+x_2^2y_1^2+y_1^2y_2^2},\ y_2' = y_2 \frac{(x_3^2+x_2^2y_1^2+y_1^2y_2^2)^{\frac{1}{2}}}{x_2^2+y_2^2}\ ; \tag{3.51}$$

$$\text{If } w = w_5,\ y_1' = y_1 \frac{(x_4^2+x_1^2y_2^2+y_1^2y_2^2)^{\frac{1}{2}}}{x_1^2+y_1^2},\ y_2' = y_2 \frac{(x_1^2+y_1^2)^{\frac{1}{2}}}{x_4^2+x_1^2y_2^2+y_1^2y_2^2}\ . \tag{3.52}$$

First assume that $w = w_2$. Then $x_1' = x_3$, $x_2' = x_2\cdot(x_2^2+y_2^2)^{-1}$, $x_3' = x_1$, $x_4' = (x_2x_3-x_1x_2^2-x_1y_2^2)\cdot(x_2^2+y_2^2)^{-1}$, $y_1' = y_1\cdot(x_2^2+y_2^2)^{\frac{1}{2}}$, $y_2' = y_2\cdot(x_2^2+y_2^2)^{-1}$. Indeed, with these values, we calculate easily that:

$$\begin{pmatrix} y_1'y_2' & x_2'y_1' & x_3' \\ & y_1' & x_1' \\ & & 1 \end{pmatrix}^{-1} \cdot w_2 \cdot \begin{pmatrix} y_1y_2 & x_2y_1 & x_3 \\ & y_1 & x_1 \\ & & 1 \end{pmatrix}$$

is an orthogonal similitude. Similarly, if $w = w_3$, we have $x_1' = x_1\cdot(x_1^2+y_1^2)^{-1}$, $x_2' = -x_4$, $x_3' = (x_1x_3+x_2y_1^2)\cdot(x_1^2+y_1^2)^{-1}$, $x_4' = -x_2$, $y_1' = y_1\cdot(x_1^2+y_1^2)^{-1}$, and $y_2' = y_2\cdot(x_1^2+y_1^2)^{\frac{1}{2}}$. As w_2, w_3 generate the Weyl group, we may now calculate $w\cdot\tau$ for all $w \in W$. In fact, using the Bruhat decomposition, we may even calculate $g\cdot\tau$ for all $g \in G$. However, we will not need such a result, for which we refer to Bump and Goldfeld [2]. (3.47-52) are sufficient for our purposes, and (3.3-8) follow.

Let us prove (3.16). Observe that:

$$I_{(\nu_1,\nu_2)}(w_1\cdot\tau) = |n_1|^{\nu_1+2\nu_2}|n_2|^{2\nu_1+\nu_2}\cdot I_{(\nu_1,\nu_2)}\left(w_1\begin{pmatrix} n_1n_2 & & \\ & n_1 & \\ & & 1\end{pmatrix}\tau\right).$$

Consequently:

$$\int_{-\infty}^{\infty}\int_{-\infty}^{\infty}\int_{-\infty}^{\infty} I_{(\nu_1,\nu_2)}\left(w_1\begin{pmatrix}1 & \xi_2 & \xi_3\\ & 1 & \xi_1\\ & & 1\end{pmatrix}\tau\right)e(-n_1\xi_1-n_2\xi_2)d\xi_1d\xi_2d\xi_3 =$$

$$|n_1|^{\nu_1+2\nu_2}|n_2|^{2\nu_1+\nu_2}\cdot\int_{-\infty}^{\infty}\int_{-\infty}^{\infty}\int_{-\infty}^{\infty} I_{(\nu_1,\nu_2)}\left(w_1\begin{pmatrix} n_1n_2 & & \\ & n_1 & \\ & & 1\end{pmatrix}\begin{pmatrix}1 & \xi_2 & \xi_3\\ & 1 & \xi_1\\ & & 1\end{pmatrix}\tau\right)$$

$$\cdot e(-n_1\xi_1-n_2\xi_2)d\xi_1d\xi_2d\xi_3 =$$

$$|n_1|^{\nu_1+2\nu_2}|n_2|^{2\nu_1+\nu_2}\cdot$$

$$\int_{-\infty}^{\infty}\int_{-\infty}^{\infty}\int_{-\infty}^{\infty} I_{(\nu_1,\nu_2)}\left(w_1\begin{pmatrix}1 & n_2\xi_2 & n_1n_2\xi_3\\ & 1 & n_1\xi_1\\ & & 1\end{pmatrix}\begin{pmatrix} n_1n_2 & & \\ & n_1 & \\ & & 1\end{pmatrix}\tau\right)$$

$$\cdot e(-n_1\xi_1-n_2\xi_2)d\xi_1d\xi_2d\xi_3 =$$

$$|n_1|^{\nu_1+2\nu_2}|n_2|^{2\nu_1+\nu_2}\cdot n_1^{-2}n_2^{-2}\cdot$$

$$\int_{-\infty}^{\infty}\int_{-\infty}^{\infty}\int_{-\infty}^{\infty} I_{(\nu_1,\nu_2)}\left(w_1\begin{pmatrix}1 & \xi_2 & \xi_3\\ & 1 & \xi_1\\ & & 1\end{pmatrix}\begin{pmatrix} n_1n_2 & & \\ & n_1 & \\ & & 1\end{pmatrix}\tau\right)$$

$$\cdot e(-\xi_1-\xi_2)d\xi_1d\xi_2d\xi_3 .$$

Multiplying both sides by $\pi^{-3\nu_1-3\nu_2+\frac{1}{2}}\Gamma(\frac{3\nu_1}{2})\Gamma(\frac{3\nu_1}{2})\Gamma(\frac{3\nu_1+3\nu_2-1}{2})$,

we obtain (3.16). We leave (3.17-22) to the reader.

The formulae (3.23-28) follow from the identity:

$$I_{(\nu_1,\nu_2)}({}^{\iota}\tau) = I_{(\nu_2,\nu_1)}(\tau) \tag{3.53}$$

We omit the proofs of (3.23-28), which are similar in style to the proof of (3.16) above.

Let us turn now to the proofs of the functional equations. We require two auxiliary integral formulae:

$$\int_{-\infty}^{\infty} (Ax^2+Bx+C)^{-\nu}\, dx = \sqrt{\pi}\; 2^{2\nu-1}\cdot(4AC-B^2)^{\frac{1}{2}-\nu}\cdot A^{\nu-1}\, \frac{\Gamma(\nu-\frac{1}{2})}{\Gamma\;(\nu)} \tag{3.54}$$

$$\int_{-\infty}^{\infty} (Ax^2+Bx+C)^{-\nu}\cdot e(-x)dx =$$

$$\frac{\pi^{\nu}}{\Gamma(\nu)}\cdot e^{\pi iB/A}\cdot(4AC-B^2)^{\frac{1}{4}-\frac{\nu}{2}}\cdot 2^{\nu+\frac{1}{2}}\cdot A^{-\frac{1}{2}}\cdot K_{\nu-\frac{1}{2}}(\pi\frac{\sqrt{4AC-B^2}}{A}) \tag{3.55}$$

In fact, (3.54) is easily evaluated by Euler's beta integral. By completing the square, (3.55) reduces to the special case $A = C = 1$, $B = 0$, which is well known (cf. Watson [38], section 6.16).

The formulae of this section follow from (3.2) if we prove them first in the special case:

$$\tau = \begin{pmatrix} y_1 y_2 & & \\ & y_1 & \\ & & 1 \end{pmatrix}$$

Thus, for the remainder of this chapter, we may assume $x_1 = x_2 = x_3 = 0$.

Let $\Omega = \{(\nu_1,\nu_2)\,|\,\mathrm{re}(\nu_1),\ \mathrm{re}(\nu_2) > \frac{1}{3}\}$. Ω is a fundamental domain for the action of W on $\mathbb{C}^2$ (cf. (2.3)). Also let:

$$\Omega_1 = \{(\nu_1,\nu_2)\,|\,\mathrm{re}(\nu_1) \neq 0 \quad \text{or} \quad \mathrm{re}(\nu_2) \neq 0\}$$

$$\Omega_2 = \{(\nu_1,\nu_2)\,|\,\mathrm{re}(\nu_1) > \tfrac{1}{3},\ \mathrm{re}(\nu_1+\nu_2) > \tfrac{2}{3}\}$$

$$\Omega_3 = \{(\nu_1,\nu_2)\,|\,\mathrm{re}(\nu_2) > \tfrac{1}{3},\ \mathrm{re}(\nu_1+\nu_2) > \tfrac{2}{3}\}\ .$$

Recall the following theorem of Hartogs:

THEOREM (Hartogs; cf. [23], Theorem 4 on p. 63). *Let U be an open set in $\mathbb{C}^{n+1}$, and let $A \subset U$ be a subset such that for each $a \in U$ there exists a neighborhood $\Delta \times D$ of a, $\Delta \subset \mathbb{C}^n$, $D \subset \mathbb{C}$ an open disk, and a positive integer p such that $\{w \in D\,|\,(z,w) \in A\}$ contains at most p points for any $z \in \Delta$. Assume that there exists a function f holomorphic on $U-A$ which is singular at every point of A; then A is a complex analytic subset of U.*

We shall use this theorem of Hartogs to prove:

THEOREM. *Let ϕ be an analytic function on Ω which admits analytic continuation to Ω_2 (resp. Ω_3) satisfying $\phi(w_2 \cdot x) = \phi(x)$ (resp. $\phi(w_3 \cdot x) = \phi(x)$) for all $x \in \Omega_2$ (resp. $x \in \Omega_3$). Then ϕ*

admits analytic continuation to all of $\mathbb{C}^2$, and $\phi(w\cdot x) = \phi(x)$ for all $x \in \mathbb{C}^2$, $w \in W$.

Proof. First we show ϕ has analytic continuation to all of Ω_1. If $x \in \Omega_1$, there exists a unique point $w\cdot x$ $(w \in W)$ lying in the closure of Ω. In this case, $w\cdot x$ is either an interior point of Ω_2 or of Ω_3, so we may define $\phi(x) = \phi(w\cdot x)$. Under our hypotheses, this function is holomorphic on Ω_1. Now let $\lambda:\mathbb{C}^2 \dashrightarrow \mathbb{C}^2$ be the map $\lambda(u,v) = (\frac{1}{3}+u+v, \frac{1}{3}+iu-iv)$. Let $X = \lambda^{-1}(\mathbb{C}^2-\Omega_1) = \{(u,v)|u = -\overline{v}\}$. Let $A \subset X$ be the singular set of $f = \phi\circ\lambda$. As the hypotheses of Hartogs' theorem are satisfied, A is a complex analytic set. As the only complex analytic subsets of X are zero-dimensional, A is zero-dimensional. However, any zero-dimensional singularity of a holomorphic function on $\mathbb{C}^2$ is removable (this follows from Theorem 2 on p. 55 of [23]). Thus f, hence ϕ, is entire.

QED.

We now address the analytic continuation of $W_{1,1}^{(\nu_1,\nu_2)}(\tau,w_1)$. If $(\nu_1,\nu_2) \in \Omega$, we have by definition:

$$W_{1,1}^{(\nu_1,\nu_2)}(\tau,w_1) = \pi^{-3\nu_1-3\nu_2+\frac{1}{2}}\Gamma(\frac{3\nu_1}{2})\Gamma(\frac{3\nu_2}{2})\Gamma(\frac{3\nu_1+3\nu_2-1}{2})\cdot I_{(\nu_1,\nu_2)}(\tau)\cdot$$

$$\int_{-\infty}^{\infty}\int_{-\infty}^{\infty}\int_{-\infty}^{\infty} (\xi_3^2+\xi_2^2y_1^2+y_1^2y_2^2)^{-3\nu_1/2}\cdot(\xi_4^2+\xi_1^2y_2^2+y_1^2y_2^2)^{-3\nu_2/2}$$

$$\cdot e(-\xi_1-\xi_2)d\xi_1 d\xi_2 d\xi_3.$$

For fixed ξ_2,ξ_3, the integral:

$$\int_{-\infty}^{\infty} (\xi_4^2+\xi_1^2y_2^2+y_1^2y_2^2)^{-3\nu_2/2}\cdot e(-\xi_1)d\xi_1$$

has the form (3.55) with $A = \xi_2^2+y_2^2$, $B = -2\xi_2\xi_3$, $C = \xi_3^2+y_1^2y_2^2$. Therefore:

$$W_{1,1}^{(\nu_1,\nu_2)}(\tau,w_1) =$$

$$2\pi^{-3\nu_1-\frac{3\nu_2}{2}+\frac{1}{2}}\cdot\Gamma(\frac{3\nu_1}{2})\Gamma(\frac{3\nu_1+3\nu_2-1}{2})\cdot y_1^{2\nu_1+\nu_2}y_2^{\frac{1}{2}+\nu_1+\frac{\nu_2}{2}}$$

$$\int_{-\infty}^{\infty}\int_{-\infty}^{\infty} (\xi_3^2+\xi_2^2y_1^2+y_1^2y_2^2)^{\frac{1}{4}-\frac{3\nu_1}{2}-\frac{3\nu_2}{4}}\cdot(\xi_2^2+y_2^2)^{-\frac{1}{2}}\cdot e(-\xi_2\xi_3/(\xi_2^2+y_2^2))$$

$$\cdot e(-\xi_2)\cdot K_{\frac{3\nu_2-1}{2}}(2\pi y_2(\xi_3^2+\xi_2^2y_1^2+y_1^2y_2^2)^{\frac{1}{2}}/(\xi_2^2+y_2^2))\, d\xi_2 d\xi_3. \qquad (3.56)$$

We now show that $W_{1,1}^{(\nu_1,\nu_2)}(\tau,w_1)$ has analytic continuation to Ω_2, and is invariant under the transformation $(\nu_1,\nu_2) \dashrightarrow w_2\cdot(\nu_1,\nu_2)$. Recall the functional equation $K_\nu = K_{-\nu}$ (cf. [38], section 3.71 (8)). We see that (3.56) is at least formally invariant under the transformation w_2 (cf. (2.5)). We must show that (3.56) converges on the interior of Ω_2, uniformly on compact sets. We will show that if $(\nu_1,\nu_2) \in \Omega_2$ does not lie in either Ω or $w_2\cdot\Omega$, then (3.56) converges on a compact neighborhood of (ν_1,ν_2). One may give also a similar, but easier proof in the case that (ν_1,ν_2) lies in Ω or $w_2\cdot\Omega$.

We show in fact that if $M > \frac{1}{3}$, $N > 0$ are chosen so that $M-N > \frac{1}{3}$, then (3.56) converges uniformly and absolutely on the set:

$$\{re(\nu_1) > M,\ \tfrac{1}{3} - N < re(\nu_2) < \tfrac{1}{3} + N\}\ . \qquad (3.57)$$

We shall require the inequality:

$$\text{If}\quad |re(\nu)| \leq A \quad\text{then}\quad |K_\nu(z)| < (\tfrac{2}{z})^A\,\Gamma(A) \qquad (3.58)$$

valid for positive numbers A and z. We shall deduce this from the well known formula:

$$K_\nu(z) = \tfrac{1}{2}\int_0^\infty e^{-\frac{z}{2}(t+\frac{1}{t})}\cdot t^\nu\cdot\frac{dt}{t}$$

This follows from [38], section 6.22 (15) after an obvious change of variables. Thus:

$$|K_\nu(z)| = \tfrac{1}{2}\left|\int_0^1 e^{-\frac{z}{2}(t+\frac{1}{t})}\cdot t^\nu\cdot\frac{dt}{t} + \int_1^\infty e^{-\frac{z}{2}(t+\frac{1}{t})}\cdot t^\nu\cdot\frac{dt}{t}\right| =$$

$$\tfrac{1}{2}\left|\int_1^\infty e^{-\frac{z}{2}(t+\frac{1}{t})}\cdot(t^\nu+t^{-\nu})\cdot\frac{dt}{t}\right| \leqq \int_1^\infty e^{-\frac{z}{2}(t+\frac{1}{t})}\cdot\tfrac{1}{2}(|t^\nu|+|t^{-\nu}|)\cdot\frac{dt}{t} \leqq$$

$$\int_1^\infty e^{-\frac{zt}{2}}\cdot t^A\cdot\frac{dt}{t} < \int_0^\infty e^{-\frac{zt}{2}}\cdot t^A\cdot\frac{dt}{t} = (\tfrac{2}{z})^A\,\Gamma(A),$$

whence (3.58). We may now prove that the integral (3.56) converges uniformly on (3.57). If A, B are two quantities depending on ν_1, ν_2, and also possibly on ξ_1,ξ_2,ξ_3, we use the notation $A \ll B$ to indicate that for the values of $\xi_1,\xi_2,\xi_3,\nu_1,\nu_2$, under consideration, there exists a constant K such that $A < K\cdot B$. By (3.58), we see that on the set (3.57), we have:

$$K_{\frac{3\nu_2-1}{2}}(2\pi y_2(\xi_3^2+\xi_2^2y_1^2+y_1^2y_2^2)^{\frac{1}{2}}/(\xi_2^2+y_2^2)) \ll \left(\frac{\sqrt{\xi_2^2+\xi_2^2y_1^2+y_1^2y_2^2}}{\xi_2^2+y_2^2}\right)^{-\frac{3N}{2}}$$

Hence the integral is dominated by:

$$\int_{-\infty}^{\infty}\int_{-\infty}^{\infty} (\xi_3^2+\xi_2^2y_1^2+y_1^2y_2^2)^{-\frac{3M}{2}}\cdot(\xi_2^2+y_2^2)^{-\frac{1}{2}+\frac{3N}{2}}\, d\xi_2 d\xi_3 \ll \quad [\text{by } (3.54)]$$

$$\int_{-\infty}^{\infty} (\xi_2^2+y_2^2)^{-\frac{3M}{2}+\frac{3N}{2}}\, d\xi_2 \ll \infty \ ,$$

again by (3.54), since $-\frac{3M}{2}+\frac{3N}{2} < -\frac{1}{2}$.

We see that (3.56) defines a holomorphic function on Ω_2 satisfying the functional equation (3.29), at least if $w = w_2$ and $(\nu_1,\nu_2) \in \Omega_2$. Similarly, we may show that $W_{1,1}^{(\nu_1,\nu_2)}(\tau, w_1)$ has analytic continuation to Ω_3 satisfying (3.29) with $w = w_3$, by proving a relation similar to (3.56), which results from (3.11) by rewriting the differential as $d\xi_1 d\xi_2 d\xi_4$, and integrating with respect to ξ_2. The global analytic continuation and functional equations (3.29) now follow from the preceding theorem.

Henceforth, we shall leave all convergence proofs to the reader.

We prove (3.30-35). Note that the meromorphic continuation of $W_{0,0}^{(\nu_1,\nu_2)}(\tau,w_0)$ to $\mathbb{C}^2$ follows from (3.10). It follows that if we prove (3.30), (3.32), and (3.34) (after analytic continuation) for

$(\nu_1,\nu_2) \in \Omega_2$, and (3.31), (3.33) and (3.35) for $(\nu_1,\nu_2) \in \Omega_3$, the meromorphic continuations of all these functions will be assured, and the functional equations will be true globally.

We now obtain the continuation of $W_{0,0}^{(\nu_1,\nu_2)}(\tau,w_3)$ and $W_{0,0}^{(\nu_1,\nu_2)}(\tau,w_5)$ to Ω_2, simultaneously proving (3.30) for $(\nu_1,\nu_2) \in \Omega_2$. Let $(\mu_1,\mu_2) \in \Omega$, and let $(\nu_1,\nu_2) = w_2\cdot(\mu_1,\mu_2)$. For fixed ξ_1, by (3.54) we have:

$$\int_{-\infty}^{\infty} (\xi_4^2+\xi_1^2y_1^2+y_1^2y_2^2)^{-3\mu_2/2}d\xi_4 = \sqrt{\pi}\, y_2^{1-3\mu_2}\cdot(\xi_1^2+y_1^2)^{\frac{1-3\mu_2}{2}}\cdot\Gamma(\tfrac{3\mu_2-1}{2})/\Gamma(\tfrac{3\mu_2}{2}).$$

Thus, (3.15) becomes:

$$W_{0,0}^{(\mu_1,\mu_2)}(\tau,w_5) = \pi^{-3\mu_1-3\mu_2+1}\cdot\Gamma(\tfrac{3\mu_1}{2})\Gamma(\tfrac{3\mu_2-1}{2})\Gamma(\tfrac{3\mu_1+3\mu_2-1}{2})$$
$$\cdot y_1^{2\mu_1+\mu_2}\cdot y_2^{1+\mu_1-\mu_2}\cdot \int_{-\infty}^{\infty} (\xi_1^2+y_1^2)^{\frac{1}{2}(1-3\mu_1-3\mu_2)}\, d\mu_1.$$

This, like the integral (3.13), converges on all of Ω_2. Furthermore, by (2.5), and the functional equation of the Riemann zeta function, we obtain (3.30) from these two integral expressions.

We prove (3.31) for $(\nu_1,\nu_2) \in \Omega_3$. Two applications of (3.54) show that the integral representation of $W_{0,0}^{(\nu_1,\nu_2)}(\tau,w_5)$ converges if $(\nu_1,\nu_2) \in \Omega_3$. Let $(\mu_1,\mu_2) \in \Omega$, and let $(\nu_1,\nu_2) = w_3\cdot(\mu_1,\mu_2)$. Let us hold ξ_1 and ξ_4 fixed, and define ξ_3 as a function of ξ_2 by $\xi_3+\xi_4 = \xi_1\xi_2$. Then by (3.54) with $A = \xi_1^2+y_1^2$, $B = -2\xi_1\xi_4$, $C = \xi_4^2+y_1^2y_2^2$, we have:

$$\int_{-\infty}^{\infty} (\xi_3^2+\xi_2^2y_1^2+y_1^2y_2^2)^{-3\mu_1/2}\, d\xi_2 =$$

$$\sqrt{\pi}\; y_1^{1-3\mu_1}\cdot(\xi_4^2+\xi_1^2y_2^2+y_1^2y_2^2)^{\frac{1-3\mu_1}{2}}\cdot(\xi_1^2+y_1^2)^{\frac{3\mu_1-2}{2}}\cdot\Gamma(\tfrac{3\mu_1-1}{2})/\Gamma(\tfrac{3\mu_1}{2}) \;.$$

Thus (3.11) becomes:

$$W_{0\,,\,0}^{(\mu_1,\mu_2)}(\tau,w_1) = \pi^{-3\mu_1-3\mu_2+1}\cdot\Gamma(\tfrac{3\mu_1-1}{2})\Gamma(\tfrac{3\mu_2}{2})\Gamma(\tfrac{3\mu_1+3\mu_2-1}{2})$$

$$\cdot y_1^{1-\mu_1+\mu_2}y_2^{\mu_1+2\mu_2}\cdot\int_{-\infty}^{\infty}\int_{-\infty}^{\infty}(\xi_1^2+y_1^2)^{\frac{3\mu_1-2}{2}}\cdot(\xi_4^2+\xi_1^2y_2^2+y_1^2y_2^2)^{\frac{1}{2}(1-3\mu_1-3\mu_2)}$$

$$\cdot d\xi_1 d\xi_4 \;.$$

This integral, as well as (3.15), is convergent on all of Ω_3. By (2.5), and the functional equation of the Riemann zeta function, we obtain (3.31).

Proof of (3.32) is similar to (3.31).

Proof of (3.33) is similar to (3.30).

We prove (3.34) for $(\nu_1,\nu_2)\in\Omega_2$. Evidently $W_{0\,,\,0}^{(\nu_1,\nu_2)}(\tau,w_0)$ has analytic continuation to Ω_2, indeed to all of $\mathbb{C}^2$. Let $(\mu_1,\mu_2)\in\Omega$, and let $(\nu_1,\nu_2) = w_2\cdot(\mu_1,\mu_2)$. By (3.54):

$$\int_{-\infty}^{\infty} (\xi_2^2+y_2^2)^{-3\mu_2/2} d\xi_2 = \sqrt{\pi}y_2^{1-3\mu_2}\cdot\Gamma(\tfrac{3\mu_2-1}{2})/\Gamma(\tfrac{3\mu_2}{2}) \;.$$

Thus (3.12) becomes:

$$W^{(\mu_1,\mu_2)}_{0,0}(\tau,w_2) =$$

$$\pi^{-3\mu_1-3\mu_2+\frac{1}{2}}\cdot\Gamma(\frac{3\mu_1}{2})\Gamma(\frac{3\mu_2-1}{2})\Gamma(\frac{3\mu_1+3\mu_2-1}{2})\cdot y_1^{2\mu_1+\mu_2}y_2^{1+\mu_1-\mu_2}.$$

By (2.5), and the functional equation of the Riemann zeta function, we obtain (3.34).

We turn now to (3.36-39). We note first that (3.40) follows from (3.13) by (3.55). (3.40) gives the meromorphic continuation of $W^{(\nu_1,\nu_2)}_{1,0}(\tau,w_3)$ to all (ν_1,ν_2). It follows that if we prove (3.36) and (3.38) for $(\nu_1,\nu_2) \in \Omega_2$, and (3.37) and (3.39) for $(\nu_1,\nu_2) \in \Omega_3$, the global meromorphic continuations and functional equations of all these functions will then be established. Indeed, (3.36) follows from an integral nearly identical to (3.56). (3.37) is similar to (3.31), (3.38) is similar to (3.30), and (3.39) follows from (3.40).

Proofs of (3.41-45) are similar to (3.36-40).

Finally, we prove (3.46). The same calculation by which we proved the absolute convergence of (3.56) on Ω_2 shows that $W^{(\nu_1,\nu_2)}_{1,1}(\tau,w_1)$ has at most polynomial growth in y_1,y_2. By Theorem 2.2, it follows that there exists a constant $c(\nu_1,\nu_2)$, independent of y_1 and y_2, such that (3.46) holds on Ω_2. Both sides of (3.46) are invariant under the action (2.3) of W, so (3.46) holds on all translates of Ω_2, i.e., the complement of the set of ν_1,ν_2 such that $re(\nu_1) = re(\nu_2) = \frac{1}{3}$. By continuity, it follows that (3.46) holds for all ν_1,ν_2.

CHAPTER IV

FOURIER EXPANSIONS OF AUTOMORPHIC FORMS

In contrast with GL(2) Fourier expansions, the correct notion of a Fourier expansion on GL(n) is by no means obvious. The essential ideas were found by Piatetski-Shapiro [24] and, independently, by Shalika [28]. These arguments were formulated in terms of global Whittaker functions on the adele group, a tool which was developed extensively by Jacquet and Langlands [14] -- see also Gelbart [4]. However, it is not hard to translate these ideas from the adele language to the classical point of view which we have taken here.

Let $\Gamma = GL(3,\mathbb{Z})$. For $\nu_1,\nu_2 \in \mathbb{C}$, an <u>automorphic form of type</u> ν_1,ν_2 is a function ϕ on $\mathcal{H}$ such that:

1. $\phi(g\cdot\tau) = \phi(\tau)$ for all $g \in \tau$, $\tau \in \mathcal{H}$;
2. We have:

$$\Delta_1\phi = \lambda\phi$$

$$\Delta_2\phi = \mu\phi$$

where $\Delta_1,\Delta_2,\lambda,\mu$ are as in Chapter 2;

3. There exist constants n_1,n_2 such that:

$$\phi\left(\begin{pmatrix} y_1y_2 & & \\ & y_1 & \\ & & 1 \end{pmatrix}\right) y_1^{n_1} y_2^{n_2}$$

is bounded on the subset of $\mathcal{H}$ determined by the inequalities $y_1, y_2 > 1$.

If furthermore:

$$\int_0^1 \int_0^1 \phi\left(\begin{pmatrix} 1 & & \xi_3 \\ & 1 & \xi_1 \\ & & 1 \end{pmatrix}\tau\right) d\xi_2 d\xi_3 = 0$$

$$\int_0^1 \int_0^1 \phi\left(\begin{pmatrix} 1 & \xi_2 & \xi_3 \\ & 1 & \\ & & 1 \end{pmatrix}\tau\right) d\xi_2 d\xi_3 = 0 ,$$

for all $\tau \in \mathcal{H}$, then ϕ is called a <u>cusp form</u>.

<u>Remark</u>: The values (ν_1, ν_2) parametrize a principal series representation of $GL(3,\mathbb{R})$ as follows: Let B be the group of upper triangular 3×3 real nonsingular matrices. B has a unique character which is trivial on the center Z, and which has the value:

$$y_1^{2\nu_1+\nu_2-1} \cdot y_2^{\nu_1+2\nu_2-1}$$

on matrices of the form:

$$\begin{pmatrix} y_1 y_2 & x_2 y_1 & x_3 \\ & y_1 & x_1 \\ & & 1 \end{pmatrix}$$

By <u>the principal series representation of</u> $GL(3,\mathbb{R})$ <u>parametrized by</u> (ν_1, ν_2), we mean the representation induced by this character, assuming, as is usually the case, that this induced representation is irreducible (cf. [5]). In order for cusp forms of type (ν_1, ν_2) to exist, this representation must be unitarizable (cf. [5], [8]). This will be the case if $re(\nu_1) = re(\nu_2) = \frac{1}{3}$. However, there exist other values of (ν_1, ν_2) which yield unitary principal series. For example, one may consider the representations induced by the complementary series representations of embedded GL(2).

It is known that cusp forms of type (ν_1, ν_2) can exist only if the principal series representation parametrized by ν_1, ν_2 is unitary. Moreover, it is conjectured that in fact cusp forms can exist only if $re(\nu_1) = re(\nu_2) = \frac{1}{3}$.

The main result of this chapter will be the Fourier expansion of an automorphic form. Let us state this as a theorem. In this statement, we shall consider only cusp forms, since for noncusp forms the situation is slightly complicated by the appearance of degenerate Whittaker functions in the Fourier expansion. However, in Chapter VII, we shall require a result which is valid for forms which are not cusp forms, and, indeed, formula (4.10) below holds for arbitrary automorphic forms.

Let Γ_∞ be the group of 3×3 upper triangular unipotent matrices with integer coefficients. Also let:

$$\Gamma^2 = \left\{ \begin{pmatrix} A & B & \\ C & D & \\ & & 1 \end{pmatrix} \middle| A,B,C,D \in \mathbb{Z}, AD-BC = \pm 1 \right\},$$

and let $\Gamma^2_\infty = \Gamma^2 \cap \Gamma_\infty$. We shall also require the subgroup Γ^2_1 of index two in Γ^2, consisting of those elements of determinant one. We shall prove:

THEOREM. *Let* ϕ *be a cusp form of type* (ν_1, ν_2). *There exist coefficients* a_{n_1,n_2}, *for positive integers* n_1, n_2, *such that:*

$$\phi(\tau) = \sum_{g\in\Gamma^2_\infty\backslash\Gamma^2} \sum_{n_1=1}^{\infty} \sum_{n_2=1}^{\infty} n_1^{-1} n_2^{-1} a_{n_1 n_2} W^{(\nu_1,\nu_2)}_{1,1}\left(\begin{pmatrix} n_1 n_2 & & \\ & n_1 & \\ & & 1 \end{pmatrix} g\tau\right)$$

We shall call the infinite array a_{n_1,n_2} the <u>matrix of Fourier coefficients</u> of ϕ.

Let us remark that for congruence subgroups, the Fourier expansions are slightly more complicated than for the full modular group. In the case of a form for a congruence subgroup, one must also sum over the cusps of the congruence subgroup of $SL(2,\mathbb{Z})$ which replaces Γ^2 in the preceding theorem. We will not consider this problem here.

This theorem was originally proved by Piatetski-Shapiro [24], in the adelic setting, on $GL(n)$. Essentially identical arguments were given by Shalika [28]. We have simply specialized the original proof by induction on n to the case $n = 3$, and translated from the adele group to $GL(3,\mathbb{R})$. Strictly speaking, the arguments of [28] are more complete, since Shalika's multiplicity-one theorem is needed at the final step in the proof.

We turn now to the proof. Since we shall require formula (4.10) below in the case of Eisenstein series, which are not cusp forms, we shall not, at first, assume that ϕ is a cusp form. Let ϕ be an automorphic form of type (ν_1,ν_2). Since ϕ is invariant under matrices of the type:

$$\begin{pmatrix} 1 & & n_2 \\ & 1 & n_1 \\ & & 1 \end{pmatrix} \qquad (n_1, n_3 \in \mathbb{Z})$$

we may write:

$$\phi(\tau) = \sum_{n_1, n_3 \in \mathbb{Z}} \phi_{n_1}^{n_3}(\tau) \tag{4.1}$$

where:

$$\phi_{n_1}^{n_3}(\tau) = \int_0^1\int_0^1 \phi\left(\begin{pmatrix} 1 & & \xi_3 \\ & 1 & \xi_1 \\ & & 1 \end{pmatrix}\tau\right) \cdot e(-n_1\xi_1 - n_3\xi_3)\,d\xi_1\,d\xi_3 \tag{4.2}$$

satisfies:

$$\phi_{n_1}^{n_3}\left(\begin{pmatrix} 1 & & \xi_3 \\ & 1 & \xi_1 \\ & & 1 \end{pmatrix}\tau\right) = e(n_1\xi_1 + n_3\xi_3)\cdot\phi_{n_1}^{n_3}(\tau) \ . \tag{4.3}$$

We show that if n_2 is an integer, then:

$$\phi_{n_1}^{n_3}\left(\begin{pmatrix} 1 & n_2 & \\ & 1 & \\ & & 1 \end{pmatrix}\tau\right) = \phi_{n_1+n_2n_3}^{n_3}(\tau) \ . \tag{4.4}$$

Indeed, the left-hand side equals:

$$\int_0^1\int_0^1 \phi\left(\begin{pmatrix} 1 & & \xi_3 \\ & 1 & \xi_1 \\ & & 1 \end{pmatrix}\begin{pmatrix} 1 & n_2 & \\ & 1 & \\ & & 1 \end{pmatrix}\tau\right) \cdot e(-n_1\xi_1 - n_3\xi_3)\,d\xi_1\,d\xi_3 =$$

$$\int_0^1\int_0^1 \phi\left(\begin{pmatrix} 1 & n_2 & \\ & 1 & \\ & & 1 \end{pmatrix}\begin{pmatrix} 1 & & \xi_3 - n_2\xi_1 \\ & 1 & \xi_1 \\ & & 1 \end{pmatrix}\tau\right) \cdot e(-(n_1+n_2n_3)\xi_1 - n_3(\xi_3 - n_2\xi_1))\,d\xi_1\,d\xi_3 .$$

Applying a change of variables, and using the fact that ϕ is automorphic, this becomes:

$$\int_0^1\int_0^1 \phi\left(\begin{pmatrix}1 & & \xi_3\\ & 1 & \xi_1\\ & & 1\end{pmatrix}\tau\right)\cdot e(-(n_1+n_2n_3)\xi_1-n_3\xi_3)d\xi_1d\xi_3 ,$$

whence (4.4).

Let $A,B,C,D,m \in \mathbb{Z}$, $AD-BC = 1$, $m > 0$. We have:

$$\phi_{mD}^{mC}(\tau) = \phi_m^0\left(\begin{pmatrix}A & B & \\ C & D & \\ & & 1\end{pmatrix}\tau\right) \tag{4.5}$$

Indeed, by (4.2) and since ϕ is automorphic, the left-hand side equals:

$$\int_0^1\int_0^1 \phi\left(\begin{pmatrix}A & B & \\ C & D & \\ & & 1\end{pmatrix}\begin{pmatrix}1 & & \xi_3\\ & 1 & \xi_1\\ & & 1\end{pmatrix}\tau\right)\cdot e(-mD\xi_1-mC\xi_3)d\xi_1d\xi_3 =$$

$$\int_0^1\int_0^1 \phi\left(\begin{pmatrix}1 & & B\xi_1+A\xi_3\\ & 1 & D\xi_1+C\xi_3\\ & & 1\end{pmatrix}\begin{pmatrix}A & B & \\ C & D & \\ & & 1\end{pmatrix}\tau\right)\cdot e(-mD\xi_1-mC\xi_3)d\xi_1d\xi_3 .$$

Changing variables, this equals:

$$\int_0^1\int_0^1 \phi\left(\begin{pmatrix}1 & & \xi_3\\ & 1 & \xi_1\\ & & 1\end{pmatrix}\begin{pmatrix}A & B & \\ C & D & \\ & & 1\end{pmatrix}\tau\right) e(-m\xi_1)d\xi_1d\xi_3 ,$$

whence (4.5)

By (4.1) and (4.5), we have:

$$\phi(\tau) = \phi_0^0(\tau) + \sum_{g\in\Gamma_\infty^2\backslash\Gamma_1^2} \sum_{m=1}^{\infty} \phi_m^0(g\cdot\tau) \tag{4.6}$$

By (4.4), ϕ_m^0 is invariant under matrices of the form:

$$\begin{pmatrix} 1 & n_2 & \\ & 1 & \\ & & 1 \end{pmatrix}$$

with $n_2 \in \mathbb{Z}$. Thus, we may write:

$$\phi_m^0(\tau) = \sum_{n=-\infty}^{\infty} \phi_{m,n}(\tau) \tag{4.7}$$

where $\phi_{m,n}$ satisfies:

$$\phi_{n_1,n_2}\left(\begin{pmatrix} 1 & \xi_2 & \xi_3 \\ & 1 & \xi_1 \\ & & 1 \end{pmatrix}\tau\right) = e(n_1\xi_1+n_2\xi_2)\cdot\phi_{n_1,n_2}(\tau) \tag{4.8}$$

We have:

$$\phi_{n_1,n_2}(\tau) = \int_0^1\int_0^1\int_0^1 \phi\left(\begin{pmatrix} 1 & \xi_2 & \xi_3 \\ & 1 & \xi_1 \\ & & 1 \end{pmatrix}\tau\right) e(-n_1\xi_1-n_2\xi_2)\,d\xi_1 d\xi_2 d\xi_3 \tag{4.9}$$

(4.6) now becomes:

$$\phi(\tau) = \sum_{n_2=-\infty}^{\infty} \phi_{0,n_2}(\tau) + \sum_{g\in\Gamma_\infty^2\backslash\Gamma_2^2} \sum_{n_1=1}^{\infty} \sum_{n_2=-\infty}^{\infty} \phi_{n_1,n_2}(g\cdot\tau) \tag{4.10}$$

Let us assume now that ϕ is a cusp form. Then ϕ_{0,n_2} and $\phi_{n_1,0}$ vanish. Since ϕ is invariant under the matrix:

$$\begin{pmatrix} -1 & & \\ & 1 & \\ & & 1 \end{pmatrix}$$

we have:

$$\phi_{n_1,n_2}\left(\begin{pmatrix} -1 & & \\ & 1 & \\ & & 1 \end{pmatrix}\tau\right) = \phi_{n_1,-n_2}(\tau) .$$

Thus, summing over Γ^2 rather than Γ_1^2, we have:

$$\phi(\tau) = \sum_{g\in\Gamma_\infty^2\backslash\Gamma^2} \sum_{n_1=1}^{\infty} \sum_{n_2=1}^{\infty} \phi_{n_1,n_2}(g\cdot\tau) \tag{4.11}$$

The theorem will thus be established if we show that, for $n_1,n_2 \neq 0$, there exists a_{n_1,n_2} such that:

$$\phi_{n_1,n_2}(\tau) = a_{n_1n_2}|n_1n_2|^{-1}.W_{1,1}^{(\nu_1,\nu_2)}\left(\begin{pmatrix} n_1n_2 & & \\ & n_1 & \\ & & 1 \end{pmatrix}\tau\right) \tag{4.12}$$

This is true whether or not ϕ is a cusp form. For the Eisenstein series, we will have an independent proof of this in Chapter VI. In general, this follows from Theorem 2.2, since the left-hand side satisfies the three conditions (1)-(3) characterizing the Whittaker function.

We have defined a_{n_1,n_2} for any nonzero n_1,n_2. We have:

$$a_{n_1,n_2} = a_{|n_1|,|n_2|} \; . \tag{4.13}$$

This may be deduced from the fact that ϕ is invariant under the four matrices:

$$\begin{pmatrix} \pm 1 & & \\ & \pm 1 & \\ & & 1 \end{pmatrix}$$

We have only reduced ϕ_{n_1,n_2} to a Whittaker function in the case $n_1 \neq 0$ and $n_2 \neq 0$. If either n_1 or n_2 vanishes, we expect ϕ_{n_1,n_2} to be a linear combination of degenerate Whittaker functions. We will see in Chapter VII that this is indeed the case for the Eisenstein series.

Finally, if ϕ is an automorphic form of type (ν_1,ν_2), we may define an automorphic form $\tilde{\phi}$ of type (ν_2,ν_1) by:

$$\tilde{\phi}(\tau) = \phi({}^\iota\tau) \tag{4.14}$$

Let a_{n_1,n_2} be the matrix of Fourier coefficients of ϕ, and let $\tilde{a}_{n_1,n_2}$ be the matrix of Fourier coefficients of $\tilde{\phi}$. We have:

$$\tilde{a}_{n_1,n_2} = a_{n_2,n_1} . \tag{4.15}$$

We leave it to the reader to deduce this from (3.24). This relation plays an essential role in the theory of L-series.

CHAPTER V

INVARIANTS OF $G_\infty \backslash G$

In this chapter, we shall associate with each orbit of $G_\infty \backslash G$ six parameters $A_1, B_1, C_1, A_2, B_2, C_2$, called the <u>invariants</u> of the orbit. We shall give a criterion, in terms of the invariants, for the orbit to contain an element of Γ. This amounts to a determination of the orbits of $\Gamma_\infty \backslash \Gamma$. Finally, we shall interpret these results in terms of the Bruhat decomposition of G.

The parametrization of this chapter was obtained independently by Bump (Dissertation, Chicago, 1982), and by Vinogradov and Takhtadzhyan [37]. It is an interesting problem to generalize this parametrization to $GL(n)$.

As usual, if $n_1, n_2, \cdots$ are integers, $(n_1, n_2, \cdots)$ denotes their greatest common divisor.

In this chapter, it will be convenient to deal only with matrices of determinant one. For this reason, we shall somewhat modify the notation of the preceding chapter. In this chapter only, we denote $G = SL(3, \mathbb{R})$, $\Gamma = SL(3, \mathbb{Z})$. G_∞ will denote the group of 3×3 upper triangular unipotent matrices, and Γ_∞ denotes $\Gamma \cap G_\infty$. Note that these groups induce the same groups of transformations on $\mathcal{H}$ as the groups denoted by the same respective symbols in the previous chapter, as the scalar matrices, which we are leaving out, act trivially.

Let:

$$\begin{pmatrix} a_{11} & a_{12} & a_{13} \\ a_{21} & a_{22} & a_{23} \\ a_{31} & a_{32} & a_{33} \end{pmatrix} \in G \tag{5.1}$$

The following six quantities depend only on the coset of the matrix (5.1) in $G_\infty \backslash G$:

$$\begin{aligned} A_1 &= -a_{31} & A_2 &= a_{22}a_{31} - a_{21}a_{32} \\ B_1 &= -a_{32} & B_2 &= a_{21}a_{33} - a_{23}a_{31} \\ C_1 &= -a_{33} & C_2 &= a_{23}a_{32} - a_{22}a_{33} \end{aligned} \tag{5.2}$$

These six quantities will be called the <u>invariants</u> of the orbit of $G_\infty \backslash G$ determined by (5.1). They are subject to one relation:

$$A_1C_2 + B_1B_2 + C_1A_2 = 0 \tag{5.3}$$

The involution (2.2) has the following effect:

$$\begin{aligned} A_1 &\longleftrightarrow A_2 \\ B_1 &\longleftrightarrow B_2 \\ C_1 &\longleftrightarrow C_2 \end{aligned} \tag{5.4}$$

This accounts for the unexpected symmetry of our results.

Our main result says that the orbits of $G_\infty \backslash G$ are in one-to-one correspondence with their invariants. We also have a characterization of the orbits of $\Gamma_\infty \backslash \Gamma$ in terms of their invariants (note that $\Gamma_\infty \backslash \Gamma$ is included injectively in $G_\infty \backslash G$).

THEOREM. (5.4) *induces a bijection of* $G_\infty \backslash G$ *onto the set* I *of all* $(A_1,B_1,C_1,A_2,B_2,C_2) \in \mathbb{R}^6$, A_1,B_1,C_1 *not all zero,* A_2,B_2,C_2 *not all zero, such that* (5.3) *is satisfied. The given orbit of* $G_\infty \backslash G$ *contains an element of* Γ *if and only if* A_1,B_1,C_1 *are coprime integers, and* A_2,B_2,C_2 *are coprime integers.*

Proof. We have already observed that (5.2) determines a well-defined map $G_\infty \backslash G \longrightarrow I$. We will show that this map is bijective.

To show that this map is injective, we show that if:

$$\begin{pmatrix} a_{11} & a_{12} & a_{13} \\ a_{21} & a_{22} & a_{23} \\ a_{31} & a_{32} & a_{33} \end{pmatrix}, \quad \begin{pmatrix} b_{11} & b_{12} & b_{13} \\ b_{21} & b_{22} & b_{23} \\ b_{31} & b_{32} & b_{33} \end{pmatrix}$$

are two matrices with the same invariants A_1,B_1,C_1,A_2,B_2,C_2, then there exist $\lambda_1,\lambda_2,\lambda_3 \in \mathbb{R}$ such that:

$$\begin{pmatrix} 1 & \lambda_2 & \lambda_3 \\ & 1 & \lambda_1 \\ & & 1 \end{pmatrix} \begin{pmatrix} a_{11} & a_{12} & a_{13} \\ a_{21} & a_{22} & a_{23} \\ a_{31} & a_{32} & a_{33} \end{pmatrix} = \begin{pmatrix} b_{11} & b_{12} & b_{13} \\ b_{21} & b_{22} & b_{23} \\ b_{31} & b_{32} & b_{33} \end{pmatrix} \tag{5.5}$$

First, we show that there exists $\lambda_1 \in \mathbb{R}$ such that:

$$\begin{pmatrix} 1 & \lambda_1 \\ & 1 \end{pmatrix} \begin{pmatrix} a_{21} & a_{22} & a_{23} \\ a_{31} & a_{32} & a_{33} \end{pmatrix} = \begin{pmatrix} b_{21} & b_{22} & b_{23} \\ b_{31} & b_{32} & b_{33} \end{pmatrix} \tag{5.6}$$

We have:

$$a_{31} = -A_1 = b_{31}$$

$$a_{32} = -B_1 = b_{32}$$

$$a_{33} = -C_1 = b_{33}$$

Also, $a_{22}a_{31} - a_{21}a_{32} = A_2 = b_{22}b_{31} - b_{21}b_{32} = b_{22}a_{31} - b_{21}a_{32}$, with similar relations for B_2, C_2. Thus:

$$a_{31}(a_{22} - b_{22}) = a_{32}(a_{21} - b_{21})$$

$$a_{32}(a_{23} - b_{23}) = a_{33}(a_{22} - b_{22})$$

$$a_{33}(a_{21} - b_{21}) = a_{31}(a_{23} - b_{23})$$

Now, a_{31}, a_{32} and a_{33} are not all zero. Suppose for example that $a_{31} \neq 0$. We may define:

$$\lambda_1 = \frac{b_{21} - a_{21}}{a_{31}}$$

The preceding relations then imply (5.6). The cases $a_{32} \neq 0$ and $a_{33} \neq 0$ are similar, so in any case, there exists λ_1 such that (5.6).

Now, we show that there exist λ_2, λ_3 such that (5.5). A_2, B_2, C_2 are not all zero. Suppose, for example, that $0 \neq A_2 = a_{22}a_{31} - a_{21}a_{32}$. The vectors (a_{21}, a_{22}) and (a_{31}, a_{32}) are thus linearly independent, so there exist λ_2, λ_3 such that:

$$(1\ \ \lambda_2\ \ \lambda_3)\begin{pmatrix} a_{11} & a_{12} \\ a_{21} & a_{22} \\ a_{31} & a_{32} \end{pmatrix} = (b_{11}\ \ b_{12})$$

Now, $-b_{13}A_2-b_{12}B_2-b_{11}C_2 = \det(b_{ij}) = 1 = \det(a_{ij}) = -a_{13}A_2-a_{12}B_2-a_{11}C_2 = -(a_{13}+\lambda_2a_{23}+\lambda_3a_{33})A_2-(a_{12}+\lambda_2a_{22}+\lambda_3a_{32})B_2-(a_{11}+\lambda_2a_{21}+\lambda_3a_{31})C_2 = -(a_{13}+\lambda_2a_{23}+\lambda_3a_{33})A_2-b_{12}B_2-b_{11}C_2$. As $A_2 \neq 0$, we have:

$$a_{13}+\lambda_2a_{23}+\lambda_3a_{33} = b_{13}$$

whence (5.5). The cases $B_2 \neq 0$ or $C_2 \neq 0$ are similar. This completes the proof that the map $G_\infty\backslash G \longrightarrow I$ is injective.

We show now that this map is surjective. Let A_1,B_1,C_1 not all zero, and A_2,B_2,C_2 not all zero, be given, and assume that (5.3) is satisfied. We may find X_1,Y_1,Z_1,X_2,Y_2,Z_2 such that:

$$A_1X_1 + B_1Y_1 + C_1Z_1 = A_2X_2 + B_2Y_2 + C_2Z_2 = 1 \tag{5.7}$$

Let:

$$\begin{array}{lll} a_{11} = -Z_2 & a_{21} = Y_1A_2 - Z_1B_2 & a_{31} = -A_1 \\ a_{12} = -Y_2 & a_{22} = Z_1C_2 - X_1A_2 & a_{32} = -B_1 \\ a_{13} = -X_2 & a_{23} = X_1B_2 - Y_1C_2 & a_{33} = -C_1 \end{array}$$

Using (5.3), one verifies the relations (5.2) for these values. Also, the determinant of (5.1) equals $A_2X_2 + B_2Y_2 + C_2Z_2 = 1$, as required. This completes the proof that the map $G_\infty \backslash G \longrightarrow I$ is surjective.

Finally, we must prove the characterization of the orbits which contain integer matrices. If (5.1) is an integer matrix, and if A_1,B_1,C_1,A_2,B_2,C_2 are given by (5.2), then evidently A_1,B_1,C_1 are coprime integers. So too are A_2,B_2,C_2, since the determinant $-a_{13}A_2-a_{12}B_2-a_{11}C_2 = 1$.

Conversely, let A_1,B_1,C_1 be coprime integers, A_2,B_2,C_2 also coprime integers, such that (5.3) is satisfied. We show that the coset parametrized by these invariants contains an integer matrix. In this case, we may find integer values X_1,Y_1,Z_1,X_2,Y_2,Z_2 satisfying (5.7). As before, we construct a matrix (5.1). It is clear that the coefficients are integers, as required.

QED.

We shall exhibit a canonical representative for each orbit. To this end, we recall the Bruhat decomposition. Let D be the group of diagonal matrices in G. If $w \in W$, denote $G_w = G_\infty w D G_\infty$, so that $G = \bigcup_w G_w$ (disjoint). The reader will easily verify the following assertions.

If $A_1 = B_1 = A_2 = B_2 = 0$, $C_1,C_2 \neq 0$, the matrix:

$$\begin{pmatrix} & & 1 \\ & 1 & \\ 1 & & \end{pmatrix}\begin{pmatrix} -1/C_2 & & \\ & C_2/C_1 & \\ & & -C_1 \end{pmatrix} \in G_{w_0} \qquad (5.8)$$

has the given invariants.

If $A_1 \neq 0,\ A_2 \neq 0,$ the matrix:

$$\begin{pmatrix} & & -1 \\ & -1 & \\ -1 & & \end{pmatrix}\begin{pmatrix} A_1 & B_1 & C_1 \\ & A_2/A_1 & -B_2/A_1 \\ & & 1/A_2 \end{pmatrix} \in G_{w_1} \tag{5.9}$$

has the given invariants.

If $A_1 = B_1 = A_2 = 0,\ C_1, B_2 \neq 0,$ the matrix:

$$\begin{pmatrix} & -1 & \\ -1 & & \\ & & -1 \end{pmatrix}\begin{pmatrix} B_2/C_1 & -C_2/C_1 & \\ & 1/B_2 & \\ & & C_1 \end{pmatrix} \in G_{w_2} \tag{5.10}$$

has the given invariants.

If $A_1 = A_2 = B_2 = 0,\ B_1, C_2 \neq 0,$ the matrix:

$$\begin{pmatrix} -1 & & \\ & & -1 \\ & -1 & \end{pmatrix}\begin{pmatrix} 1/C_2 & & \\ & B_1 & C_1 \\ & & C_2/B_1 \end{pmatrix} \in G_{w_3} \tag{5.11}$$

has the given invariants.

If $A_2 = 0,\ A_1, B_2 \neq 0,$ the matrix:

$$\begin{pmatrix} & 1 & \\ & & 1 \\ 1 & & \end{pmatrix}\begin{pmatrix} -A_1 & -B_1 & -C_1 \\ & -1/B_2 & \\ & & B_2/A_1 \end{pmatrix} \in G_{w_4} \tag{5.12}$$

has the given invariants.

If $A_1 = 0$, $B_1, A_2 \neq 0$, the matrix:

$$\begin{pmatrix} & & 1 \\ 1 & & \\ & 1 & \end{pmatrix}\begin{pmatrix} A_2/B_1 & & \\ & -B_1 & -C_1 \\ & & -1/A_2 \end{pmatrix} \in G_{w_5} \tag{5.13}$$

has the given invariants.

CHAPTER VI

RAMANUJAN SUMS ON GL(3)

In 1918, Ramanujan [27] considered the following sums:

$$c_q(n) = \sum_{\substack{p \bmod q \\ (p,q)=1}} e(\frac{np}{q})$$

which bear his name. His investigation dealt with interpreting, exactly or approximately, the number of representations of a number by a quadratic form as a divisor sum. Today, we regard such results as relations between Eisenstein series and theta series, although these relations were not explicit in Ramanujan's work, where the Eisenstein series appeared only as Lambert series.

Ramanujan found the following relation:

$$\zeta(s)\sum_{q=1}^{\infty} c_q(n)q^{-s} = \sigma_{1-s}(n) \tag{6.1}$$

where:

$$\sigma_\nu(n) = \sum_{d|n} d^\nu$$

Let us recall a proof of this fact. We have:

$$\sum_{d|q} c_d(n) = \sum_{p \bmod q} e\left(\frac{np}{q}\right) = \begin{cases} q & \text{if } q|n \\ 0 & \text{otherwise.} \end{cases}$$

Consequently:

$$\zeta(s)\sum_{q=1}^{\infty} c_q(n)q^{-s} = \sum_{q=1}^{\infty}\{\sum_{d|q} c_d(n)\}\, q^{-s} = \sum_{q|n} q^{1-s} = \sigma_{1-s}(n) .$$

On GL(3), also, we find that Ramanujan sums occur in the theory of Eisenstein series. In this chapter, however, we shall consider these sums for their own interest, without regard for their connection with the Eisenstein series.

We shall state first the results, then give the proofs.

If A_1, B_1, C_1, A_2, and B_2 are given, $A_1, A_2 \neq 0$, let us define C_2 by requiring that:

$$A_1C_2 + B_1B_2 + C_1A_2 = 0 \qquad (6.2)$$

Now, for fixed A_1, A_2, B_1, B_2, let us denote by $\Lambda_{A_1,A_2}(B_1,B_2)$ the number of C_1 mod A_1 (or, equivalently, the number of C_2 mod A_2) such that C_1 and C_2 are both integers, and $(A_1,B_1,C_1) = (A_2,B_2,C_2)=1$, where (A_1,B_1,C_1) denotes the greatest common divisor of A_1, B_1 and C_1. We define the following Ramanujan sum:

$$R_{A_1,A_2}(n_1,n_2) = \sum_{\substack{B_1 \bmod A_1 \\ B_2 \bmod A_2}} \Lambda_{A_1,A_2}(B_1,B_2)e(\frac{n_1B_1}{A_1} + \frac{n_2B_2}{A_2}) \qquad (6.3)$$

Let us also explain the GL(3) analog of the divisor sums which occur in Ramanujan's formula (6.1). Firstly, let us reconsider the classical divisor sums as characters of finite-dimensional representations of GL(2). The divisor sum is multiplicative:

If $(n,n') = 1$, then $\sigma_a(nn') = \sigma_a(n)\sigma_a(n')$.

Thus, it is sufficient to consider the case of a prime power $\sigma_a(p^k)$. Then:

$$\sigma_a(p) = 1 + p^a + \ldots + p^{ak} = S_k(1,p^a)$$

where S_k is the k-th <u>Schur polynomial</u>:

$$S_k(\alpha,\beta) = \alpha^k + \alpha^{k-1}\beta + \ldots + \beta^k = \frac{\begin{vmatrix} \alpha^{k+1} & \beta^{k+1} \\ 1 & 1 \end{vmatrix}}{\begin{vmatrix} \alpha & \beta \\ 1 & 1 \end{vmatrix}} .$$

The Schur polynomials have an interpretation as characters of the irreducible finite-dimensional representations of GL(2). Let V be the standard two-dimensional representation of $GL(2,\mathbb{C})$. Any (algebraic) finite-dimensional irreducible representation of $GL(2,\mathbb{C})$ is then a symmetric power $\vee^k(V)$, times a power of the determinant. If α, β are the eigenvalues of $A \in GL(2,\mathbb{C})$, and if χ_k is the character of the representation $\vee^k V$, then we have:

$$\chi_k(A) = S_k(\alpha, \beta)$$

We may now generalize the divisor sums to $GL(3)$. Firstly, let us recall the irreducible algebraic characters on $GL(3,\mathbb{C})$; for these, we refer to Weyl [39], Dieudonne and Carroll [2b], Humphreys [10], and Macdonald [22C]. If k_1, k_2 are nonnegative integers, these parametrize a character χ_{k_1,k_2} of $GL(3)$, namely, if α, β and γ are the eigenvalues of $A \in GL(3)$, then:

$$\chi_{k_1,k_2}(A) = S_{k_1,k_2}(\alpha,\beta,\gamma) \tag{6.4}$$

where:

$$S_{k_1,k_2}(\alpha,\beta,\gamma) = \frac{\begin{vmatrix} \alpha^{k_1+k_2+2} & \beta^{k_1+k_2+2} & \gamma^{k_1+k_2+2} \\ \alpha^{k_1+1} & \beta^{k_1+1} & \gamma^{k_1+1} \\ 1 & 1 & 1 \end{vmatrix}}{\begin{vmatrix} \alpha^2 & \beta^2 & \gamma^2 \\ \alpha & \beta & \gamma \\ 1 & 1 & 1 \end{vmatrix}} \tag{6.5}$$

is a <u>Schur polynomial</u>. Any algebraic character of GL(n) is the product of one of these characters by a power of the determinant.

Note that the numerator in the definition is an alternating polynomial function of α, β, and γ, and as such must be divisible by the Vandermonde determinant in the denominator. Thus, S_{k_1,k_2} is a symmetric polynomial in α, β, and γ.

EXERCISE: Compute $S_{2,3}(\alpha,\beta,\gamma)$.

We will prove an analog of Ramanujan's formula. We will show that:

$$\zeta(\nu_1)\zeta(\nu_2)\zeta(\nu_1+\nu_2-1)\sum_{A_1=1}^{\infty}\sum_{A_2=1}^{\infty} R_{A_1,A_2}(n_1,n_2)A_1^{-\nu_1}A_2^{-\nu_2} =$$

$$\sigma_{1-\nu_1,1-\nu_2}(n_1,n_2) \tag{6.6}$$

where $\sigma_{\nu_1,\nu_2}(n_1,n_2)$ is an analog of the divisor sum $\sigma_\nu(n)$. Specifically, define $\sigma_{\nu_1,\nu_2}(n_1,n_2)$ by requiring that:

If $(n_1n_2, n_1'n_2') = 1$, then $\sigma_{\nu_1,\nu_2}(n_1n_2', n_2n_2') =$

$$\sigma_{\nu_1,\nu_2}(n_1n_2)\sigma_{\nu_1,\nu_2}(n_1',n_2') \tag{6.7}$$

and, if p is a prime, denoting:

$$\alpha = p^{\nu_1}, \quad \beta = p^{\nu_2}$$

we require that:

$$\sigma_{\nu_1,\nu_2}(p^{k_1},p^{k_2}) = \beta^{-k_1} S_{k_1,k_2}(1,\beta,\alpha\beta) \tag{6.8}$$

For example, we have:

$$\sigma_{\nu_1,\nu_2}(p,p) = \beta^{-1} S_{1,1}(1,\beta,\alpha\beta) =$$

$$\left\{\begin{array}{lllll} & 1 & + \; p^{\nu_1} & & \\ + & p^{\nu_2} & + \; 2p^{\nu_1+\nu_2} & + & p^{2\nu_1+\nu_2} \\ & & + \; p^{\nu_1+2\nu_2} & + & p^{2\nu_1+2\nu_2} \; . \end{array}\right.$$

The coefficients in this array will be familiar to physicists.

To study the sums $R_{A_1,A_2}(n_1,n_2)$, it is convenient to simultaneously study the auxiliary sums $r_{A_1,A_2}(n_1,n_2)$ and $s_{A_1,A_2}(n_1,n_2)$ defined by:

$$r_{A_1,A_2}(n_1,n_2) = \sum_{\substack{a_1 | A_1 \\ a_2 | A_2}} R_{a_1,a_2}(n_1,n_2) \tag{6.9}$$

and:

$$s_{A_1,A_2}(n_1,n_2) = \sum_{d|(A_1,A_2)} d\,r_{\frac{A_1}{d},\frac{A_2}{d}}(n_1,n_2) \tag{6.10}$$

The first properties which we need are multiplicativity conditions:

If $(A_1A_2,A_1'A_2') = 1$, then $R_{A_1A_1',A_2A_2'}(n_1,n_2) =$

$$R_{A_1A_2}(n_1,n_2)\cdot R_{A_1',A_2'}(n_1,n_2)\ . \tag{6.11}$$

The same precise property is valid for the sums r and s:

If $(A_1A_2,A_1'A_2') = 1$, then $r_{A_1A_1',A_2A_2'}(n_1,n_2) =$

$$r_{A_1,A_2}(n_1,n_2)\cdot r_{A_1',A_2'}(n_1,n_2)\ ; \tag{6.12}$$

If $(A_1A_2,A_1'A_2') = 1$, then $s_{A_1A_1',A_2A_2'}(n_1,n_2) =$

$$s_{A_1,A_2}(n_1,n_2)\cdot s_{A_1',A_2'}(n_1,n_2)\ . \tag{6.13}$$

Thus, each sum can be reduced to a product of sums where A_1 and A_2 are powers of the same prime p. Also, n_1 and n_2 may be assumed to be powers of p, because:

If $(m_1m_2, A_1A_2) = 1$, then $R_{A_1,A_2}(m_1n_1, m_2n_2) = R_{A_1,A_2}(n_1,n_2)$.

(6.14)

If $(m_1m_2, A_1A_2) = 1$, then $r_{A_1,A_2}(m_1n_1, m_2n_2) = r_{A_1,A_2}(n_1,n_2)$.

(6.15)

If $(m_1m_2, A_1A_2) = 1$, then $s_{A_1,A_2}(m_1n_1, m_2n_2) = s_{A_1,A_2}(n_1,n_2)$.

(6.16)

Furthermore, in the case where A_1, A_2, n_1, n_2 are all powers of the same prime p, we may evaluate $s_{A_1,A_2}(n_1,n_2)$ explicitly. If M_1, M_2, m_1, m_2 are nonnegative integers, define:

$$M_{m_1,m_2}(M_1,M_2) = \begin{cases} M_0 & \text{if } M_0 > 0 ; \\ 0 & \text{otherwise,} \end{cases}$$

where:

$$M_0 = \min(m_1,M_1) + \min(m_2,M_2) + 1 - \max(m_1,m_2) .$$

We will obtain the following expression for the s sum in the case A_1, A_2, n_1, n_2 are powers of the same prime p:

$$s_{p^{m_1},p^{m_2}}(p^{M_1},p^{M_2}) = M_{m_1,m_2}(M_1,M_2)p^{m_1+m_2} \tag{6.17}$$

Since the R and r sums may be reconstructed from the s sums by Möbius inversion, this amounts to a complete determination of the Ramanujan sums.

Let us mention two more formulae which we shall require. The first is a simpler expression for the r sums:

$$r_{A_1,A_2}(n_1,n_2) = (A_1,A_2) \sum_{\substack{B_1 \bmod A_1 \\ B_2 \bmod A_2 \\ (A_1,A_2)|B_1B_2}} e\left(\frac{n_1B_1}{A_1}+\frac{n_2B_2}{A_2}\right) \tag{6.18}$$

Also, let us mention that the coefficients $M_{m_1,m_2}(M_1,M_2)$ may be interpreted as the weight multiplicities in the representation of $GL(3,\mathbb{C})$ with character χ_{M_1,M_2}. This means that the coefficients occur as coefficients in the Schur polynomials. Specifically,

$$M_{m_1,m_2}(M_1,M_2) = 0 \qquad \text{Unless} \quad 0 \leq m_1,m_2 \leq M_1+M_2 \tag{6.19}$$

and:

$$\beta^{-M_1}S_{M_1,M_2}(1,\beta,\alpha\beta) = \sum_{m_1=1}^{M_1+M_2}\sum_{m_2=1}^{M_1+M_2} M_{m_1,m_2}(M_1,M_2)\alpha^{m_1}\beta^{m_2} \tag{6.20}$$

We turn now to the proofs. We shall prove (6.6) last of all.

Let us prove (6.11). Let $a_1 = A_1A_1'$, $a_2 = A_2A_2'$, $b_1 = B_1A_1'+B_1'A_1$, and $b_2 = B_2A_2'+B_2'A_2$. Note that as B_1, B_2, B_1', B_2' run through the residue classes mod A_1, A_2, A_1', A_2', respectively, b_1 and b_2 run through the residue classes mod a_1, a_2 respectively. Let $L_{A_1,A_2}(B_1,B_2)$ denote the set of C_1 mod A_1 (or, equivalently, the number of C_2 mod A_2) such that C_1, C_2 are both integers, and $(A_1,B_1,C_1) = A_2,B_2,C_2) = 1$. Thus $\Lambda_{A_1,A_2}(B_1,B_2)$ is the cardinality of $L_{A_1,A_2}(B_1B_2)$ by definition. We will define a bijection between the Cartesian product $L_{A_1,A_2}(B_1,B_2) \times L_{A_1',A_2'}(B_1',B_2')$ and $L_{a_1,a_2}(b_1,b_2)$. This will prove $\Lambda_{a_1,a_2}(b_1,b_2) = \Lambda_{A_1,A_2}(B_1,B_2) \cdot \Lambda_{A_1',A_2'}(B_1',B_2')$. The bijection is defined as follows. Since by hypothesis $(A_1A_2, A_1'A_2') = 1$, find p,q so that $pA_1A_2+qA_1'A_2' = 1$. Now, given C_1, C_2 (resp. C_1', C_2') representing an element of $L_{A_1,A_2}(B_1,B_2)$ (resp. $L_{A_1',A_2'}(B_1',B_2')$), we define c_1, c_2 representing an element of $L_{a_1,a_2}(b_1,b_2)$ as follows:

$$\begin{aligned} c_1 &= A_1C_1' + A_1'C_1 - pA_1^2B_1'B_2 - qA_1'^2B_1B_2' \\ c_2 &= A_2C_2' + A_2'C_2 - pA_2^2B_2'B_1 - qA_2'^2B_2B_1' \end{aligned} \tag{6.21}$$

One calculates that $a_1c_2 + b_1b_2 + c_1a_2 = 0$. We leave it to the reader to show that $(a_1,b_1,c_1) = (a_2,b_2,c_2) = 1$. Thus, we have defined a map from $L_{A_1,A_2}(B_1,B_2) \times L_{A_1',A_2'}(B_1',B_2')$ to $L_{a_1,a_2}(b_1,b_2)$. It may be seen that this map is a bijection, whence:

$$\Lambda_{a_1,a_2}(b_1,b_2) = \Lambda_{A_1,A_2}(B_1,B_2) \cdot \Lambda_{A_1',A_2'}(B_1',B_2') . \tag{6.22}$$

On the other hand, we have:

$$e(\frac{n_1b_1}{a_1}+\frac{n_2b_2}{a_2}) = e(\frac{n_1B_1}{A_1}+\frac{n_2B_2}{A_2})e(\frac{n_1B_1'}{A_1'}+\frac{n_2B_2'}{A_2'}) \tag{6.23}$$

Combining (6.22) and (6.23) and summing over B_1,B_2,B_1',B_2', we obtain (6.11).

<u>Remark</u>: Another way of proving (6.11), avoiding the parametrization of c_1,c_2 by (6.21) is as follows. First one proves (6.18), defining r_{A_1,A_2} without specific reference to C_1,C_2. Then, one proves (6.12) without needing (6.21). (6.11) follows from (6.12) by Möbius inversion. We have chosen the particular proof given here because for the exponential sums associated with certain other Eisenstein series, the technique of parametrization as in (6.22) is definitely required. Although these questions are beyond the scope of this book, we prefer to use the more general method.

(6.12) and (6.13) are easily deduced from (6.11). Let us prove (6.14). We define a bijection of $L_{A_1,A_2}(B_1,B_2)$ with $L_{A_1,A_2}(m_1B_1,m_2B_2)$. Given C_1,C_2 representing an element of $L_{A_1,A_2}(B_1,B_2)$, we associate $m_1m_2C_1$ and $m_1m_2C_2$, representing an element of $L_{A_1,A_2}(m_1B_1,m_2B_2)$. This shows that:

$$\Lambda_{A_1,A_2}(m_1B_1,m_2B_2) = \Lambda_{A_1,A_2}(B_1,B_2) \tag{6.24}$$

Now, as B_1,B_2 run over the residue classes mod A_1,A_2 respectively, so do m_1B_1 and m_2B_2. Consequently, we have:

$$R_{A_1,A_2}(m_1n_1,m_2n_2) = \sum_{\substack{B_1 \bmod A_1 \\ B_2 \bmod A_2}} \Lambda_{A_1,A_2}(B_1,B_2)e\left(\frac{m_1n_1B_1}{A_1}+\frac{m_2n_2B_2}{A_2}\right) =$$

$$\sum_{\substack{m_1B_1 \bmod A_1 \\ m_2B_2 \bmod A_2}} \Lambda_{A_1,A_2}(m_1B_1,m_2B_2)e\left(\frac{n_1m_1B_1}{A_1}+\frac{n_2m_2B_2}{A_2}\right) = R_{A_1,A_2}(n_1n_2) ,$$

whence (6.14).

(6.15) and (6.16) follow easily from (6.14).

Let us prove (6.18). We have:

$$\sum_{\substack{a_1|A_1 \\ a_2|A_2}} R_{a_1,a_2}(n_1,n_2) = \sum_{\substack{a_1|A_1 \\ a_2|A_2}} \sum_{\substack{b_1 \bmod a_1 \\ b_2 \bmod a_2}} \sum_{\substack{c_1 \bmod a_1 \\ c_2 \bmod a_2 \\ a_1c_2+b_1b_2+c_1a_2=0 \\ (a_1,b_1,c_1)=1 \\ (a_2,b_2,c_2)=1}} e\left(\frac{n_1b_1}{a_1}+\frac{n_2b_2}{a_2}\right)$$

In this sum, find d_1,d_2 so that $A_1 = a_1d_1$, $A_2 = a_2d_2$, and let $B_1 = b_1d_1$, $B_2 = b_2d_2$, $C_1 = c_1d_1$, $C_2 = c_2d_2$. Our sum then equals:

$$\sum_{\substack{d_1|A_1 \\ d_2|A_2}} \sum_{\substack{B_1 \bmod A_1 \\ B_2 \bmod A_2}} \sum_{\substack{C_1 \bmod A_1 \\ C_2 \bmod A_2 \\ A_1C_2+B_1B_2+C_1A_2=0 \\ (A_1,B_1,C_1)=d_1 \\ (A_2,B_2,C_2)=d_2}} e\left(\frac{n_1B_1}{A_1}+\frac{n_2B_2}{A_2}\right) =$$

$$\sum_{\substack{B_1 \bmod A_1 \\ B_2 \bmod A_2}} \sum_{\substack{C_1 \bmod A_1 \\ C_2 \bmod A_2 \\ A_1C_2+B_1B_2+C_1A_2=0}} e\left(\frac{n_1B_1}{A_1}+\frac{n_2B_2}{A_2}\right) .$$

Now, given B_1, B_2, the number of $C_1 \bmod A_1$, $C_2 \bmod A_2$ such that (6.2) is satisfied is:

$$\begin{cases} (A_1,A_2) & \text{if } (A_1,A_2) \mid B_1B_2 ; \\ 0 & \text{otherwise.} \end{cases}$$

(6.18) now follows.

We prove (6.17). By symmetry, we may assume that $m_1 \geqq m_2$. Furthermore, we may assume that $M_1 \leqq m_1$, since if $M_1 > m_1$, neither the left nor the right side of the equation is changed if M_1 is replaced by m_1; similarly, we may assume that $M_2 \leqq m_2$. We will make these three assumptions. By (6.18):

$$r_{p^{m_1},p^{m_2}}(p^{M_1},p^{M_2}) = p^{m_2}\sum_{k=0}^{m_2} \sum_{\substack{B_1 \bmod p^{m_1} \\ \operatorname{ord}(B_1)\geqq m_2-k}} \sum_{\substack{B_2 \bmod p^{m_2} \\ \operatorname{ord}(B_2)=k}} e(B_1p^{M_1-m_1}+B_2p^{M_2-m_2}),$$

where $\operatorname{ord}(B) = e$, the largest integer such that $p^e \mid B$. Thus:

$$s_{p^{m_1},p^{m_2}}(p^{M_1},p^{M_2}) = \sum_{h=0}^{m_2} p^h r_{p^{m_1-h},p^{m_2-h}}(p^{M_1},p^{M_2})$$

$$= p^{m_2}\sum_{h=0}^{m_2} \sum_{k=0}^{m_2-h} \sum_{\substack{B_1 \bmod p^{m_1-h} \\ \operatorname{ord}(B_1)\geqq m_2-h-k}} \sum_{\substack{B_2 \bmod p^{m_2-h} \\ \operatorname{ord}(B_2) = k}} e(B_1p^{M_1-m_1+h}+B_2p^{M_2-m_2+h})$$

Let us replace B_1, B_2 in this sum by $b_1 = p^h B_1$, $b_2 = p^h B_2$. Our previous sum equals:

$$p^{m_2} \sum_{k=0}^{m_2} \sum_{h=0}^{m_2-k} \sum_{\substack{b_1 \bmod p^{m_1} \\ \operatorname{ord}(b_1) \geqq m_2-k}} \sum_{\substack{b_2 \bmod p^{m_2} \\ \operatorname{ord}(b_2)=k+h}} e(b_1 p^{M_1-m_1} + b_2 p^{M_2-m_2})$$

$$= p^{m_2} \sum_{k=0}^{m_2} \sum_{\substack{b_1 \bmod p^{m_1} \\ \operatorname{ord}(b_1) \geqq m_2-k}} \sum_{\substack{b_2 \bmod p^{m_2} \\ \operatorname{ord}(b_2) \geqq k}} e(b_1 p^{M_1-m_1} + b_2 p^{M_2-m_2})$$

We have:

$$\sum_{\substack{b \bmod p^m \\ \operatorname{ord}(b) \geqq k}} e(bp^{M-m}) = \begin{cases} p^{m-k} & \text{if } M+k \geqq m \,; \\ 0 & \text{otherwise.} \end{cases}$$

Thus, in our previous sum, the contribution of a particular value of k is:

$$\begin{array}{ll} p^{m_2} p^{m_1-m_2+k} p^{m_2-k} = p^{m_1+m_2} & \text{if } m_2-M_2 \leqq k \leqq M_1-m_1+m_2 \,; \\ 0 & \text{otherwise.} \end{array}$$

Summing, we obtain (6.17).

We prove (6.20). For notational convenience, let us fix M_1, M_2, and denote $M_{m_1,m_2}(M_1,M_2)$ as $M(m_1,m_2)$.

Let $N(m_1,m_2) = M(m_1,m_2) - M(m_1-1,m_2)$. We leave it to the reader to verify that:

$$N(m_1,m_2) = \begin{cases} 1 & \text{if } 0 \leq m_1 \leq M_1 \\ & \text{and } 0 \leq m_2-m_1 \leq M_2 \\ -1 & \text{if } M_1+1 \leq m_1 \leq M_1+M_2+1 \\ & \text{and } 0 \leq m_2-m_1+M_1+1 \leq M_2 \\ 0 & \text{otherwise.} \end{cases}$$

Consequently, we have:

$$M(m_1,m_2) - M(m_1-1,m_2) - M(m_1,m_2-1)$$
$$+ M(m_1-2,m_2-1) + M(m_1-1,m_2-2) - M(m_1-2,m_2-2)$$

$$= N(m_1,m_2) - N(m_1,m_2-1) - N(m_1-1,m_2-1) + N(m_1-1,m_2-2) =$$

$$\begin{cases} 1 & \text{if } m_1 = 0, & m_2 = 0, \\ -1 & \text{if } m_1 = M_1+1, & m_2 = 0, \\ -1 & \text{if } m_1 = 0, & m_2 = M_2+1, \\ 1 & \text{if } m_1 = M_1+M_2+1, & m_2 = M_2+1, \\ 1 & \text{if } m_1 = M_1+1, & m_2 = M_1+M_2+1, \\ -1 & \text{if } m_1 = M_1+M_2+1, & m_2 = M_1+M_2+1. \\ 0 & \text{otherwise.} \end{cases}$$

This is the coefficient of $\alpha^{m_1}\beta^{m_2}$ in:

$$(1-\alpha-\beta+\alpha\beta^2+\alpha^2\beta-\alpha^2\beta^2) \cdot \sum_{m_1,m_2} M(m_1,m_2)\alpha^{m_1}\beta^{m_2}$$

Thus:

$$\sum_{m_1=1}^{M_1+M_2} \sum_{m_2=1}^{M_1+M_2} M(m_1,m_2)\alpha^{m_1}\beta^{m_2}$$

$$= \frac{1-a^{M_1+1}-b^{M_2+1}+a^{M_1+1}b^{M_1+M_2+1}+a^{M_1+M_2+1}b^{M_2+1}-a^{M_1+M_2+1}b^{M_1+M_2+1}}{1-a-b+ab^2+a^2b-a^2b^2}$$

$$= \beta^{-M_1}S(1,\beta,\alpha\beta) ,$$

whence (6.20).

We may now prove (6.6). Indeed, we will prove that:

$$\zeta(\nu_1)\zeta(\nu_2)\zeta(\nu_1+\nu_2-1) \sum_{A_1=1}^{\infty} \sum_{A_2=1}^{\infty} R_{A_1,A_2}(n_1,n_2)A_1^{-\nu_1}A_2^{-\nu_2}$$

$$= \zeta(\nu_1+\nu_2-1) \sum_{A_1=1}^{\infty} \sum_{A_2=1}^{\infty} r_{A_1,A_2}(n_1,n_2)A_1^{-\nu_1}A_2^{-\nu_2}$$

$$= \sum_{A_1=1}^{\infty} \sum_{A_2=1}^{\infty} s_{A_1,A_2}(n_1,n_2)A_1^{-\nu_1}A_2^{-\nu_2}$$

$$= \sigma_{1-\nu_1,1-\nu_2}(n_1,n_2) \qquad (6.25)$$

The first two equalities follow from the definitions of r and s. As for the third, let:

$$n_i = \prod_p p^{e_i(p)} \qquad (i = 1,2)$$

be the factorizations of n_1, n_2 into primes. By (6.13) and (6.16),

$$\sum_{A_1} \sum_{A_2} s_{A_1,A_2}(n_1,n_2) A_1^{-\nu_1} A_2^{-\nu_2} =$$

$$\prod_p \sum_{m_1=0}^{\infty} \sum_{m_2=0}^{\infty} s_{p^{m_1},p^{m_2}}(p^{e_1(p)}, p^{e_2(p)}) p^{-m_1\nu_1 - m_2\nu_2}$$

By (6.17) and (6.20),

$$\sum_{m_1=0}^{\infty} \sum_{m_2=0}^{\infty} s_{p^{m_1},p^{m_2}}(p^{e_1}, p^{e_2}) p^{-m_1\nu_1 - m_2\nu_2}$$

$$= \sum_{m_1=0}^{\infty} \sum_{m_2=0}^{\infty} M_{m_1,m_2}(e_1,e_2) \alpha^{m_1} \beta^{m_2}$$

$$= \beta^{-e_1} S_{e_1,e_2}(1,\beta,\alpha\beta) = \sigma_{1-\nu_1,1-\nu_2}(p^{e_1}, p^{e_2})$$

where

$$\alpha = p^{1-\nu_1} \qquad \beta = p^{1-\nu_2} .$$

(6.25), and thence (6.6), now follow from (6.7).

CHAPTER VII
EISENSTEIN SERIES

In this chapter, we shall consider the "minimal parabolic" Eisenstein series. In addition to the usual group-theoretic definition, we shall give an expression for the Eisenstein series, analogous to an Epstein zeta function, as the sum of values of a quadratic form, over the orthogonal pairs of points in two dual lattices in $\mathbb{R}^3$. We shall obtain the Fourier expansions, and the analytic continuation and functional equations, and we shall discuss the structure of the polar divisor.

The expression for the Eisenstein series, analogous to an Epstein zeta function, is the starting point for the investigation of Bump and Goldfeld [2] on the Kronecker limit formula for cubic fields. This expression depends on the result of Chapter V parametrizing $G_\infty \backslash G$. It is unclear how this expression will generalize to $GL(n)$.

The most important fact about the Eisenstein series, the analytic continuation and functional equations, was obtained in complete generality for arbitrary reductive groups by Langlands [20]; see also Harish-Chandra [8]. Langland's method, strongly influenced by ideas of Selberg, obtains first the analytic continuation and functional equations of the "constant term" of the Eisenstein series, relying on methods of spectral theory to show that the Eisenstein series themselves must then satisfy the same functional equations. A different method is available when one has the complete Fourier expansions, where one may consider each Fourier term separately. We shall follow this latter method.

The Fourier expansions of $GL(3)$ minimal parabolic Eisenstein series were obtained independently by Bump (dissertation, 1982) and by

Vinogradov and Takhtadzhyan [37]. Imai and Terras [11] also obtained Fourier expansions, for more general Eisenstein series on GL(3), including the maximal parabolic Eisenstein series (see below). However, their Fourier expansions are different from ours, being in terms of an abelian subgroup of the parabolic, rather than the complete parabolic, and using in place of the Whittaker function a so-called K-Bessel function. Thus, it is not trivial to relate the expansions of their paper to ours. A fourth paper, due to Proskurin [26], is in Russian. We have been unable to examine this paper.

The Fourier coefficients of the Eisenstein series are essentially the generalized divisor sums introduced in the previous chapter. It was noted there that these sums are multiplicative, and, when the indexing parameters are both powers of the same prime p, the sum may be interpreted as a special value of a character of an irreducible representation of GL(3). This surprising phenomenon is related to a theorem of Shintani [31], identifying values of the p-adic Whittaker functions on GL(n) as character values.

It is also possible to consider "maximal parabolic" Eisenstein series, constructed from GL(2) cusp forms; for these, we refer to Imai and Terras [11]. It is also possible to consider maximal parabolic Eisenstein series with the GL(2) cusp form replaced by the constant function; for these, see Friedberg [3]. The latter Eisenstein series are actually residues of the minimal parabolic Eisenstein series.

We shall give first the statements, then the proofs, of the results of this chapter.

If $re(\nu_1), re(\nu_2) > \frac{2}{3}$, $\tau \in \mathcal{H}$, the Eisenstein series:

$$E_{(\nu_1,\nu_2)}(\tau) = \sum_{g \in \Gamma_\infty \backslash \Gamma} I_{(\nu_1,\nu_2)}(g\cdot\tau)$$

is well-defined, and is absolutely convergent. The theorem of Chapter V allows us to rewrite this result. If we associate with the matrix g invariants A_1,B_1,C_1,A_2,B_2,C_2 as in (6.2), we will show that:

$$I_{(\nu_1,\nu_2)}(g\cdot\tau) = I_{(\nu_1,\nu_2)}(\tau)\cdot[(A_1x_3+B_1x_1+C_1)^2+(A_1x_2+B_1)^2y_1^2+A_1^2y_1^2y_2^2]^{-3\nu_1/2}\cdot$$
$$[(A_2x_4-B_2x_2+C_2)^2+(A_2x_1-B_2)^2y_2^2+A_2^2y_1^2y_2^2]^{-3\nu_2/2} \qquad (7.1)$$

where x_1,x_2,x_3,x_4,y_1,y_2 are the coordinates introduced in Chapter II. Thus, by the theorem of Chapter V, we have:

$$E_{(\nu_1,\nu_2)}(\tau) =$$
$$I_{(\nu_1,\nu_2)}(\tau)\cdot\sum_{\substack{(A_1,B_1,C_1)=1\\(A_2,B_2,C_2)=1\\A_1C_2+B_1B_2+C_1A_2=0}}[(A_1x_3+B_1x_1+C_1)^2+(A_1x_2+B_1)^2y_1^2+A_1^2y_1^2y_2^2]^{-3\nu_1/2}\cdot[(A_2x_4-B_2x_2+C_2)^2+(A_2x_1-B_2)^2y_2^2+A_2^2y_1^2y_2^2]^{-3\nu_2/2}. \qquad (7.2)$$

The following renormalization of the Eisenstein series is somewhat more convenient. Let:

$$G_{(\nu_1,\nu_2)}(\tau) = \tfrac{1}{4}\pi^{-3\nu_1-3\nu_2+\frac{1}{2}}\cdot\Gamma(\frac{3\nu_1}{2})\Gamma(\frac{3\nu_2}{2})\Gamma(\frac{3\nu_1+3\nu_2-1}{2})\cdot$$
$$\zeta(3\nu_1)\zeta(3\nu_2)\zeta(3\nu_1+3\nu_2-1)\cdot E_{(\nu_1,\nu_2)}(\tau) \qquad (7.3)$$

We combine this with (7.2). The factors $\zeta(3\nu_1)$, $\zeta(3\nu_2)$ may be absorbed into the sum simply by dropping the condition that $(A_1,B_1,C_1) = (A_2,B_2,C_2) = 1$. Thus, we obtain the following expression, which is homogeneous in the sense that it is the sum over the nonzero lattice points in a homogeneous real algebraic set:

$$G_{(\nu_1,\nu_2)}(\tau) = \tfrac{1}{4}\pi^{-3\nu_1-3\nu_2+\frac{1}{2}} \cdot \Gamma\left(\frac{3\nu_1}{2}\right)\Gamma\left(\frac{3\nu_2}{2}\right)\Gamma\left(\frac{3\nu_1+3\nu_2-1}{2}\right) \cdot \zeta(3\nu_1+3\nu_2-1) \cdot$$

$$I_{(\nu_1,\nu_2)}(\tau) \sum_{A_1C_2+B_1B_2+C_1A_2=0}{}' \left[(A_1x_3+B_1x_1+C_1)^2+(A_1x_2+B_1)^2y_1^2+A_1^2y_1^2y_2^2\right]^{-3\nu_1/2} \cdot$$

$$\left[(A_2x_4-B_2x_2+C_2)^2+(A_2x_1-B_2)^2y_2^2+A_2^2y_1^2y_2^2\right]^{-3\nu_2/2} \tag{7.4}$$

Here $\sum'$ denotes summation over all $A_1,B_1,C_1 \in \mathbb{Z}$, not all zero, and over all $A_2,B_2,C_2 \in \mathbb{Z}$, not all zero, subject to the relation $A_1C_2 + B_1B_2 + C_1A_2 = 0$.

We consider now the interpretation of the Eisenstein series as an "Epstein zeta function." If S is an $n \times n$ real symmetric matrix, and if $\xi \in \mathbb{R}^n$, let $S[\xi]$ denote ${}^t\xi \cdot S \cdot \xi \in \mathbb{R}$. Let:

$$S_\tau = y_1^{-\frac{4}{3}} y_2^{-\frac{2}{3}} \begin{pmatrix} y_1y_2 & y_1x_2 & x_3 \\ & y_1 & x_1 \\ & & 1 \end{pmatrix} \begin{pmatrix} y_1y_2 & & \\ y_1x_2 & y_1 & \\ x_3 & x_1 & 1 \end{pmatrix}$$

so that $\det(S_\tau) = 1$. Noting that:

$$S_\tau\begin{bmatrix}A_1\\B_1\\C_1\end{bmatrix} = [(A_1x_3+B_1x_1+C_1)^2 + (A_1x_2+B_1)^2y_1^2+A_1^2y_1^2y_2^2]\cdot y_1^{-\frac{4}{3}}y_2^{-\frac{2}{3}}$$

$$S_\tau^{-1}\begin{bmatrix}C_2\\B_2\\A_2\end{bmatrix} = [(A_2x_4-B_2x_2+C_2)^2 + (A_2x_1-B_2)^2y_2^2+A_2y_1y_2]\cdot y_1^{-\frac{2}{3}}y_2^{-\frac{4}{3}}$$

we see that:

$$G_{(\nu_1,\nu_2)}(\tau) =$$

$$\tfrac{1}{4}\pi^{-3\nu_1-3\nu_2+\frac{1}{2}}\Gamma(\tfrac{3\nu_1}{2})\Gamma(\tfrac{3\nu_2}{2})\Gamma(\tfrac{3\nu_1+3\nu_2-1}{2})\zeta(3\nu_1+3\nu_2-1)\cdot$$

$$\sum_{\substack{0\neq\xi,\eta\in\mathbb{Z}^3\\ \langle\xi,\eta\rangle=0}} S_\tau[\xi]^{-\frac{3\nu_1}{2}}\ S_\tau^{-1}[\eta]^{-\frac{3\nu_2}{2}} \tag{7.5}$$

where the sum is over nonzero orthogonal integer vectors:

$$\xi = \begin{pmatrix}A_1\\B_1\\C_1\end{pmatrix} \qquad \eta = \begin{pmatrix}C_2\\B_2\\A_2\end{pmatrix}.$$

The most important fact about Eisenstein series is that they have analytic continuation and functional equations:

THEOREM 7.1. $G_{(\nu_1,\nu_2)}(\tau)$ *has meromorphic continuation to all values of* ν_1,ν_2, *and is invariant under the action* (2.5) *of the Weyl group. The polar divisor of* $G_{(\nu_1,\nu_2)}(\tau)$ *consists of the six lines:*

$$\begin{aligned} \nu_1 &= 0 \text{ or } \tfrac{2}{3}\,; \\ \nu_2 &= 0 \text{ or } \tfrac{2}{3}\,; \\ 1-\nu_1-\nu_2 &= 0 \text{ or } \tfrac{2}{3}\,. \end{aligned}$$

This is a special case of the general theorem of Langlands [20].

We will obtain the Fourier expansions of the Eisenstein series. We will use the notations of Chapter VI for the generalized divisor sums and Schur polynomials.

THEOREM 7.2. *The Eisenstein series* $G_{(\nu_1,\nu_2)}(\tau)$ *has a Fourier expansion* (4.10). *If* $n_1,n_2 \neq 0$,

$$\phi_{n_1,n_2} = |n_1n_2|^{-1}\, a_{n_1,n_2} W^{(\nu_1,\nu_2)}_{1,1}\left(\begin{pmatrix} n_1n_2 & & \\ & n_1 & \\ & & 1 \end{pmatrix}\tau\right)$$

where:

$$a_{n_1,n_2} = |n_1|^{\nu_1+2\nu_2-1}\,|n_2|^{2\nu_1+\nu_2-1}\,\sigma_{1-3\nu_2,1-3\nu_1}(|n_1|,|n_2|)$$

in terms of the divisor sums of Chapter VI. The degenerate terms $\phi_{0,0}$, $\phi_{n_1,0}$ *and* ϕ_{0,n_2} *are given by* (7.6-8) *below.*

We have:

$$\phi_{0,0}(\tau) = \zeta(3\nu_1)\zeta(3\nu_2)\zeta(3\nu_1+3\nu_2-1)\cdot W^{(\nu_1,\nu_2)}_{0\,,\,0}(\tau,w_0) +$$

$$\zeta(3\nu_1)\zeta(3\nu_2-1)\zeta(3\nu_1+3\nu_2-1)\cdot W^{(\nu_1,\nu_2)}_{0\,,\,0}(\tau,w_2) +$$

$$\zeta(3\nu_1-1)\zeta(3\nu_2)\zeta(3\nu_1+3\nu_2-2)\cdot W^{(\nu_1,\nu_2)}_{0\,,\,0}(\tau,w_4) +$$

$$\zeta(3\nu_1-1)\zeta(3\nu_2-1)\zeta(3\nu_1+3\nu_2-2)\cdot W^{(\nu_1,\nu_2)}_{0\,,\,0}(\tau,w_1)$$

$$\zeta(3\nu_1)\zeta(3\nu_2-1)\zeta(3\nu_1+3\nu_2-2)\cdot W^{(\nu_1,\nu_2)}_{0\,,\,0}(\tau,w_5) +$$

$$\zeta(3\nu_1-1)\zeta(3\nu_2)\zeta(3\nu_1+3\nu_2-1)\cdot W^{(\nu_1,\nu_2)}_{0\,,\,0}(\tau,w_3) \qquad (7.6)$$

If $n_1 \neq 0$, we have:

$$\phi_{n_1,0}(\tau) = \zeta(3\nu_2)\zeta(3\nu_1+3\nu_2-1)|n_1|^{\nu_1-\nu_2-1}\sigma_{1-3\nu_1}(|n_1|)W_{1,0}\left(\begin{pmatrix} n_1 & & \\ & n_1 & \\ & & 1\end{pmatrix}\tau, w_3\right)$$

$$+\ \zeta(3\nu_1)\zeta(3\nu_2-1)|n_1|^{\nu_1+2\nu_2-2}\sigma_{2-3\nu_1-3\nu_2}(|n_1|)W_{1,0}\left(\begin{pmatrix} n_1 & & \\ & n_1 & \\ & & 1\end{pmatrix}\tau, w_5\right)$$

$$+\ \zeta(3\nu_1-1)\zeta(3\nu_1+3\nu_2-2)|n_1|^{\nu_1+2\nu_2-2}\sigma_{1-3\nu_2}(|n_1|)W_{1,0}\left(\begin{pmatrix} n_1 & & \\ & n_1 & \\ & & 1\end{pmatrix}\tau, w_1\right)$$

(7.7)

And, if $n_2 \neq 0$, we have:

$$\phi_{0,n_2}(\tau) =$$

$$\zeta(3\nu_1)\zeta(3\nu_1+3\nu_2-1)|n_2|^{-\nu_1+\nu_2-1}\sigma_{1-3\nu_2}(|n_2|)W_{0,1}\left(\begin{pmatrix} n_2 & & \\ & 1 & \\ & & 1\end{pmatrix}\tau, w_2\right)$$

$$+\ \zeta(3\nu_2)\zeta(3\nu_1-1)|n_2|^{2\nu_1+\nu_2-2}\sigma_{2-3\nu_1-3\nu_2}(|n_2|)W_{0,1}\left(\begin{pmatrix} n_2 & & \\ & 1 & \\ & & 1\end{pmatrix}\tau, w_4\right)$$

$$+\ \zeta(3\nu_2-1)\zeta(3\nu_1+3\nu_2-2)|n_2|^{2\nu_1+\nu_2-2}\sigma_{1-3\nu_1}(|n_2|)W_{0,1}\left(\begin{pmatrix} n_2 & & \\ & 1 & \\ & & 1\end{pmatrix}\tau, w_1\right)$$

(7.8)

If n_1, n_2 are powers of the same prime p, we have an expression for a_{n_1,n_2} in terms of the Schur polynomials S_{k_1,k_2} introduced in Chapter VI. As in Chapter II, let:

$$\alpha = -\nu_1 - 2\nu_2 + 1$$

$$\beta = -\nu_1 + \nu_2$$

$$\gamma = 2\nu_1 + \nu_2 - 1$$

Then, in the notation of Chapter VI, we have:

$$a_{p^{k_1}, p^{k_2}} = S_{k_1,k_2}(p^\alpha, p^\beta, p^\gamma) = \chi_{k_1,k_2}\left(\begin{pmatrix} p^\alpha & & \\ & p^\beta & \\ & & p^\gamma \end{pmatrix}\right) \tag{7.9}$$

The occurrence of characters of irreducible representations of $GL(n,\mathbb{C})$ in the Fourier expansions of automorphic forms is of course no coincidence. A p-adic result of Shintani [31] shows that this phenomenon may be expected to generalize (at least) to $GL(n)$.

Shintani's result is an explicit formula for the Whittaker function of a class one principal series representation of $GL(n)$ over a p-adic field; essentially, the values of the p-adic Whittaker function are given by the irreducible characters of $GL(n,\mathbb{C})$, applied to a semisimple conjugacy class determined by the given principal series representation of $GL(n)$. This has implications for the Fourier expansions of automorphic forms because, if one considers the p-localization of an automorphic representation, the p-adic Whittaker function essentially gives the p-local Fourier coefficients of the form. For a discussion, in the $GL(2)$ case, of the connection between the Whittaker functions and Fourier expansion, cf. Gelbart [4], sections 6B and 6C, and also Lemma 3.6.

From this point of view, Shintani's result gives the Fourier coefficients of a Hecke eigenform in terms of the corresponding representation of the Hecke algebra. Let ϕ be an automorphic form on GL(3), as defined in Chapter IV. (Because Shintani considers only class one representations, if we consider more generally forms with respect to congruence subgroups, his result only applies directly to coefficients a_{n_1,n_2} with n_1,n_2 prime to the conductor.) If ϕ is a Hecke eigenform, then for each p, ϕ determines a representation of the local Hecke algebra H_p. The structure of H_p is known; and in particular (cf. Langlands [21]), such a representation of H_p determines a semisimple conjugacy class A_p in $GL(3,\mathbb{C})$.

Now, according to Shintani's theorem, the Fourier coefficient $a_{p^{k_1},p^{k_2}}$ is essentially the trace of A_p on the representation module for $GL(n,\mathbb{C})$ associated with the character $\chi_{p^{k_1},p^{k_2}}$. Thus, the formula (7.9) is an expected consequence of Shintani's theorem.

The Fourier coefficients are multiplicative. We have:

$$\text{If } (n_1n_2,n_1'n_2') = 1, \text{ then } a_{n_1n_1',n_2n_2'} = a_{n_1,n_2}a_{n_1',n_2'} . \tag{7.10}$$

and:

$$a_{1,1} = 1 . \tag{7.11}$$

We turn now to the proofs.

The absolute convergence of the Eisenstein series $E_{(\nu_1,\nu_2)}(\tau)$ when $re(\nu_1)$, $re(\nu_2) > \frac{2}{3}$ follows from a criterion of Godement, for which we refer to Baily [1], Lemma 4 on p. 194, in the case $G = SL(3,\mathbb{Z})$, P minimal parabolic.

(7.1) follows from combining the canonical forms (5.5-10) with (3.4-9). For example, if g lies in the coset of $G_\infty\backslash G$ parametrized by (5.6), we have:

$$I_{(\nu_1,\nu_2)}(g\cdot\tau) = I_{(\nu_1,\nu_2)}\left(w_1\cdot\begin{pmatrix} A_1 & B_1 & C_1 \\ & A_2/A_1 & -B_2/A_1 \\ & & 1/A_2 \end{pmatrix}\cdot\tau\right)$$

If τ is given by coordinates x_1,x_2,x_3,x_4,y_1,y_2 as in Chapter II, we have:

$$\begin{pmatrix} A_1 & B_1 & C_1 \\ & A_2/A_1 & -B_2/A_1 \\ & & 1/A_2 \end{pmatrix}\cdot\tau\cdot\begin{pmatrix} A_2 & & \\ & A_2 & \\ & & A_2 \end{pmatrix} = \begin{pmatrix} \eta_1\eta_2 & \xi_2\eta_1 & \xi_3 \\ & \eta_1 & \xi_1 \\ & & 1 \end{pmatrix}$$

where:

$$\begin{aligned} \xi_1 &= (A_2/A_1)(A_2x_1-B_2) & \eta_1 &= A_2^2y_1/A_1 \\ \xi_2 &= (A_1/A_2)(A_1x_2+B_1) & \eta_2 &= A_1^2y_2/A_2 \\ \xi_3 &= A_2\cdot(A_1x_3+B_1x_1+C_1) . \end{aligned}$$

Using (5.3), we see that $\xi_4 = \xi_1\xi_2 - \xi_3$ has the value:

$$\xi_4 = A_1 \cdot (A_2 x_4 - B_2 x_2 + C_2)$$

By (3.5), we obtain (7.1). The other cases are handled similarly.

(7.2-5) follow from (7.1).

We turn now to the Fourier expansions. Let us prove Theorem 7.2. First note that since each Fourier term satisfies (4.8), we may assume that:

$$\tau = \begin{pmatrix} y_1 y_2 & & \\ & y_1 & \\ & & 1 \end{pmatrix};$$

i.e., that $x_1 = x_2 = x_3 = x_4 = 0$. We make this assumption.

Assume that $n_1, n_2 \neq 0$. We have:

$$\phi_{n_1,n_2}(\tau) =$$

$$\int_0^1 \int_0^1 \int_0^1 G_{(\nu_1,\nu_2)}\left(\begin{pmatrix} 1 & \xi_2 & \xi_3 \\ & 1 & \xi_1 \\ & & 1 \end{pmatrix} \tau\right) \cdot e(-n_1\xi_1 - n_2\xi_2)\, d\xi_1 d\xi_2 d\xi_3 =$$

$$\tfrac{1}{4}\pi^{-3\nu_1-3\nu_2+\frac{1}{2}} \Gamma\left(\frac{3\nu_1}{2}\right)\Gamma\left(\frac{3\nu_2}{2}\right)\Gamma\left(\frac{3\nu_1+3\nu_2-1}{2}\right) \cdot I_{(\nu_1,\nu_2)}(\tau) \cdot \zeta(3\nu_1+3\nu_2-1) \cdot$$

$$\sum_{A_1C_2+B_1B_2+C_1A_2=0}{}' \int_0^1 \int_0^1 \int_0^1 [(A_1\xi_3+B_1\xi_1+C_1)^2 + (A_1\xi_2+B_1)^2 y_1^2 + A_1^2 y_1^2 y_2^2]^{-3\nu_1/2} \cdot$$

$$[(A_2\xi_4-B_2\xi_2+C_2)^2 + (A_2\xi_1-B_2)^2 y_2^2 + A_2 y_1^2 y_2^2]^{-3\nu_2/2} \cdot e(-n_1\xi_1-n_2\xi_2)\, d\xi_1 d\xi_2 d\xi_3 .$$

The terms where $A_1 = 0$ or $A_2 = 0$ vanish. For example, if $A_1 = 0$ but $A_2 \neq 0$ and $B_1 \neq 0$, the integrand still involves ξ_2, but after the transformation:

$$\xi_1 \longrightarrow \xi_1 + \frac{B_2}{A_2} \qquad \xi_2 \longrightarrow \xi_2$$

$$\xi_3 \longrightarrow \xi_3 \qquad \xi_4 \longrightarrow \xi_4 + \frac{B_2}{A_2}\xi_2$$

which preserves the relation (2.1), the integrand (but for the exponential) becomes independent of ξ_2. As $n_2 \neq 0$, the term vanishes. In all other cases, too, where $A_1 = 0$ or $A_2 = 0$, the term vanishes. We may thus assume, in the summation that $A_1 \neq 0$, $A_2 \neq 0$; indeed, by dropping the factor $\frac{1}{4}$, we may assume that $A_1, A_2 > 0$. Thus:

$$\phi_{n_1,n_2}(\tau) =$$

$$\pi^{-3\nu_1-3\nu_2+\frac{1}{2}}\Gamma(\tfrac{3\nu_1}{2})\Gamma(\tfrac{3\nu_2}{2})\Gamma(\tfrac{3\nu_1+3\nu_2-1}{2})\cdot I_{(\nu_1,\nu_2)}(\tau)\cdot\zeta(3\nu_1+3\nu_2-1)\cdot$$

$$\sum_{A_1=1}^{\infty}\sum_{A_2=1}^{\infty}\sum_{\substack{B_1,C_1 \bmod A_1\\ B_2,C_2 \bmod A_2\\ A_1C_2+B_1B_2+C_1A_2=0}}\int_{-\infty}^{\infty}\int_{-\infty}^{\infty}\int_{-\infty}^{\infty}[(A_1\xi_3+B_1\xi_1+C_1)^2+(A_1\xi_2+B_1)^2y_1^2+A_1^2y_1^2y_2^2]^{-3\nu_1/2}\cdot$$

$$[(A_2\xi_4-B_2\xi_2+C_2)^2+(A_2\xi_1-B_2)^2y_2^2+A_2^2y_1^2y_2^2]^{-3\nu_2/2}\cdot$$

$$e(-n_1\xi_1-n_2\xi_2)\,d\xi_1\,d\xi_2\,d\xi_3\ .$$

Now let us apply the change of variables:

$$\xi_1 \longrightarrow \xi_1 + \frac{B_2}{A_2} \qquad \xi_2 \longrightarrow \xi_2 - \frac{B_1}{A_1}$$

$$\xi_3 \longrightarrow \xi_3 - \frac{B_1\xi_1}{A_1} + \frac{C_2}{A_2} \qquad \xi_4 \longrightarrow \xi_4 + \frac{B_2\xi_2}{A_2} + \frac{C_1}{A_1}$$

By (5.3), this transformation preserves the relation (2.1). Also by (5.3), the integral simplifies to:

$$\pi^{-3\nu_1-3\nu_2+\frac{1}{2}}\Gamma\left(\frac{3\nu_1}{2}\right)\Gamma\left(\frac{3\nu_2}{2}\right)\Gamma\left(\frac{3\nu_1+3\nu_2-1}{2}\right)\cdot I_{(\nu_1,\nu_2)}(\tau)\cdot$$

$$\zeta(3\nu_1+3\nu_2-1)\sum_{A_1=1}^{\infty}\sum_{A_2=1}^{\infty}A_1^{-3\nu_1}A_2^{-3\nu_2}\cdot\sum_{\substack{B_1,C_1 \bmod A_1\\ B_2,C_2 \bmod A_2\\ A_1C_2+B_1B_2+C_1A_2=0}} e\left(\frac{n_2B_1}{A_1}-\frac{n_1B_2}{A_2}\right)\cdot$$

$$\int_{-\infty}^{\infty}\int_{-\infty}^{\infty}\int_{-\infty}^{\infty}(\xi_3^2+\xi_2^2y_1^2+y_1^2y_2^2)^{-3\nu_1/2}(\xi_4^2+\xi_1^2y_2^2+y_1^2y_2^2)^{-3\nu_2/2}\cdot$$

$$e(-n_1\xi_1-n_2\xi_2)d\xi_1d\xi_2d\xi_3\ .$$

By (3.11) and (3.16), we see that:

$$a_{n_1,n_2} = |n_1|^{\nu_1+2\nu_2-1}|n_2|^{2\nu_1+\nu_2-1}\zeta(3\nu_1+3\nu_2-1)\cdot$$

$$\sum_{A_1=1}^{\infty}\sum_{A_2=1}^{\infty}A_1^{-3\nu_1}A_2^{-3\nu_2}\sum_{\substack{B_1,C_1 \bmod A_1\\ B_2,C_2 \bmod A_2\\ A_1C_2+B_1B_2+C_1A_2=0}} e\left(\frac{n_2B_1}{A_1}-\frac{n_1B_2}{A_2}\right)\ .$$

Now, given A_1,A_2,B_1,B_2, there exists $C_1 \bmod A_1$ or, equivalently, $C_2 \bmod A_2$ such that $A_1C_2 + B_1B_2 + C_1A_2 = 0$ if and only if $(A_1,A_2)|B_1B_2$.

In this case, there exist (A_1,A_2) such solutions. Thus, by (6.18):

$$\sum_{\substack{B_1,C_1 \bmod A_1 \\ B_2,C_2 \bmod A_2 \\ A_1C_2+B_1B_2+C_1A_2=0}} e\left(\frac{n_2B_1}{A_1} - \frac{n_1B_2}{A_2}\right) =$$

$$(A_1,A_2) \sum_{\substack{B_1 \bmod A_1 \\ B_2 \bmod A_2 \\ (A_1,A_2)|B_1B_2}} e\left(\frac{n_2B_1}{A_1} - \frac{n_1B_2}{A_2}\right) =$$

$$r_{A_2,A_1}(-n_1,n_2) = r_{A_2,A_1}(|n_1|,|n_2|) \, .$$

By (6.25), we now obtain:

$$a_{n_1,n_2} = |n_1|^{\nu_1+2\nu_2-1}|n_2|^{2\nu_1+\nu_2-1}\sigma_{1-3\nu_2,1-3\nu_1}(|n_1|,|n_2|) \, ,$$

as required for Theorem 7.2.

Before evaluating the degenerate terms (7.6-8), let us point out that (7.9-11) follow from this formula. As for (7.9), we have, by (6.8):

$$a_{p^{k_1},p^{k_2}}$$
$$= p^{k_1(\nu_1+2\nu_2-1)+k_2(2\nu_1+\nu_2-1)}p^{-k_1(1-3\nu_1)}S_{k_1,k_2}(1,p^{1-3\nu_1},p^{2-3\nu_1-3\nu_2})$$
$$= p^{(2k_1+k_2)(2\nu_1+\nu_2-1)}S_{k_1,k_2}(1,p^{1-3\nu_1},p^{2-3\nu_1-3\nu_2}) \, .$$

Since S_{k_1,k_2} is homogeneous of degree $2k_1 + k_2$, we obtain (7.9). The multiplicativity assertions (7.10) and (7.11) follow from the corresponding multiplicativity (6.7) for the generalized divisor sums.

We turn now to (7.6). First we consider the contribution of the terms $A_1 \neq 0$, $A_2 \neq 0$. By a calculation similar to the nondegenerate case just considered, the contribution of these terms to $\phi_{0,0}$ is:

$$\zeta(3\nu_1+3\nu_2-1) \sum_{A_1=1}^{\infty} \sum_{A_2=1}^{\infty} A_1^{-3\nu_1} A_2^{-3\nu_2} \left\{ \sum_{\substack{B_1,C_1 \bmod A_1 \\ B_2,C_2 \bmod A_2 \\ A_1C_2+B_1B_2+C_1A_2=0}} 1 \right\} W_{0,0}^{(\nu_1,\nu_2)}(\tau,w_1) =$$

$$\zeta(3\nu_1+3\nu_2-1) \prod_p \sum_{k_1=0}^{\infty} \sum_{k_2=0}^{\infty} r_{p^{k_1},p^{k_2}}(p^{k_1},p^{k_2}) p^{-3k_1\nu_1-3k_2\nu_2} . W_{0,0}^{(\nu_1,\nu_2)}(\tau,w_1).$$

We show then that:

$$\prod_p \sum_{k_1=0}^{\infty} \sum_{k_2=0}^{\infty} r_{p^{k_1},p^{k_2}}(p^{k_1},p^{k_2})\, p^{-3k_1\nu_1-3k_2\nu_2} =$$

$$\zeta(3\nu_1-1)\zeta(3\nu_2-1)\zeta(3\nu_1+3\nu_2-2)\zeta(3\nu_1+3\nu_2-1)^{-1}$$

By (6.10) and Möbius inversion, we have:

$$r_{A_1,A_2}(n_1,n_2) = \sum_{d|(A_1,A_2)} \mu(d) d s_{\frac{A_1}{d},\frac{A_2}{d}}(n_1,n_2) \tag{7.12}$$

In particular:

$$r_{p^{k_1},p^{k_2}}(n_1,n_2) =$$

$$\begin{cases} s_{p^{k_1},p^{k_2}}(n_1,n_2) - ps_{p^{k_1-1},p^{k_2-1}}(n_1,n_2) & \text{if } k_1,k_2 > 0 \,; \\ s_{p^{k_1},p^{k_2}}(n_1,n_2) & \text{otherwise.} \end{cases} \tag{7.13}$$

Thus, by (6.17):

$$r_{p^{k_1},p^{k_2}}(p^{k_1},p^{k_2}) =$$

$$\begin{cases} (\min(k_1,k_2)+1)\cdot p^{k_1+k_2} - \min(k_1,k_2)\cdot p^{k_1+k_2-1} & \text{if } k_1,k_2 > 0 \,; \\ p^{k_1+k_2} & \text{otherwise.} \end{cases}$$

Thus, the coefficient of $p^{-3k_1\nu_1-3k_2\nu_2}$ in:

$$(1-p\cdot p^{-3\nu_1})(1-p\cdot p^{-3\nu_2}) \sum_{k_1=0}^{\infty} \sum_{k_2=0}^{\infty} r_{p^{k_1},p^{k_2}}(p^{k_1},p^{k_2})\cdot p^{-3k_1\nu_1-3k_2\nu_2}$$

equals:

$$\begin{cases} 0 & \text{if } k_1 \neq k_2 \,; \\ 1 & \text{if } k_1 = k_2 = 0 \,; \\ p^{2k}-p^{2k-1} & \text{if } k_1 = k_2 > 0 \,. \end{cases}$$

Thus our previous infinite product equals:

$$\prod_p (1-p^{1-3\nu_1})^{-1}\cdot(1-p^{1-3\nu_2})^{-1}\cdot(1+(p^2-p)p^{-3\nu_1-3\nu_2}+(p^4-p^3)p^{-6\nu_1-6\nu_2}+\dots)=$$

$$\prod_p (1-p^{1-3\nu_1})^{-1}\cdot(1-p^{1-3\nu_2})^{-1}\cdot(1-p^{1-3\nu_1-3\nu_2})\cdot(1-p^{2-3\nu_1-3\nu_2})^{-1},$$

as required. Hence the contribution of the terms in question is:

$$\zeta(3\nu_1-1)\zeta(3\nu_2-1)\zeta(3\nu_1+3\nu_2-2)\cdot W_{0,0}^{(\nu_1,\nu_2)}(\tau,w_1)$$

Next we sum the terms with $A_1 = B_1 = 0$, $A_2 = B_2 = 0$. By absorbing the factor $\frac{1}{4}$ into the sum, we may assume that $C_1 > 0$, $C_2 > 0$, and the sum of these terms equals:

$$\zeta(3\nu_1+3\nu_2-1) \sum_{C_1=1}^{\infty} \sum_{C_2=1}^{\infty} C_1^{-3\nu_1} C_2^{-3\nu_2}\cdot W_{0,0}^{(\nu_1,\nu_2)}(\tau,w_0)$$

Hence the contribution of the terms in question is:

$$\zeta(3\nu_1)\zeta(3\nu_2)\zeta(3\nu_1+3\nu_2-1)\cdot W_{0,0}^{(\nu_1,\nu_2)}(\tau,w_0)$$

Next we sum the terms with $A_1 = B_1 = A_2 = 0$, $C_1 \neq 0$, $B_2 \neq 0$. Similarly to the previous calculations, we obtain:

$$\zeta(3\nu_1+3\nu_2-1) \sum_{C_1=1}^{\infty} \sum_{B_2=1}^{\infty} \sum_{C_2 \bmod B_2} C_1^{-3\nu_1} B_2^{-3\nu_2} \cdot W_{0\,,\,0}^{(\nu_1,\nu_2)}(\tau, w_2) =$$

$$\zeta(3\nu_1+3\nu_2-1) \sum_{C_1=1}^{\infty} \sum_{B_2=1}^{\infty} C_1^{-3\nu_1} B_2^{1-3\nu_2} \cdot W_{0\,,\,0}^{(\nu_1,\nu_2)}(\tau, w_2) .$$

Hence the contribution of the terms in question is:

$$\zeta(3\nu_1)\zeta(3\nu_2-1)\zeta(3\nu_1+3\nu_2-1) \cdot W_{0\,,\,0}^{(\nu_1,\nu_2)}(\tau, w_2) .$$

Similarly, the contribution of the terms where $A_1 = A_2 = B_2 = 0$, $B_1 \neq 0$, $C_2 \neq 0$ is:

$$\zeta(3\nu_1-1)\zeta(3\nu_2)\zeta(3\nu_1+3\nu_2-1) \cdot W_{0\,,\,0}^{(\nu_1,\nu_2)}(\tau, w_3)$$

Next we sum the terms where $A_2 = 0$, $A_1 \neq 0$, $B_2 \neq 0$. Similarly to the previous calculations, these terms contribute:

$$\zeta(3\nu_1+3\nu_2-1) \sum_{A_1=1}^{\infty} \sum_{A_2=1}^{\infty} \sum_{\substack{B_1, C_1 \bmod A_1 \\ C_2 \bmod B_2 \\ A_1C_2+B_1B_2=0}} A_1^{-3\nu_1} B_2^{-3\nu_2} \cdot W_{0\,,\,0}^{(\nu_1,\nu_2)}(\tau, w_4) .$$

Of course, the number of $C_1 \bmod A_1$ is A_1, and the number of $B_1 \bmod A_1$, $C_2 \bmod B_2$ such that $A_1C_2 + B_1B_2 = 0$ is (A_1,B_2). Thus:

$$\sum_{A_1=1}^{\infty} \sum_{B_1=1}^{\infty} \sum_{\substack{B_1, C_1 \bmod A_1 \\ C_2 \bmod B_2 \\ A_1C_2+B_1B_2=0}} A_1^{-3\nu_1} B_2^{-3\nu_2} =$$

$$\sum_{A_1=1}^{\infty} \sum_{B_1=1}^{\infty} (A_1,B_2) \cdot A_1^{1-3\nu_1} B_2^{-3\nu_2} =$$

$$\prod_p \sum_{k_1=0}^{\infty} \sum_{k_2=0}^{\infty} (p^{k_1}, p^{k_2}) \cdot p^{k_1(1-3\nu_1)-3k_2\nu_2} =$$

$$\prod_p (1-p^{2-3\nu_1-3\nu_2})^{-1} \cdot \left(\sum_{k_1=0}^{\infty} p^{k_1(1-3\nu_1)} + \sum_{k_2=0}^{\infty} p^{-3k_2\nu_2} - 1 \right) =$$

$$\prod_p (1-p^{2-3\nu_1-3\nu_2})^{-1} \cdot \left((1-p^{1-3\nu_1})^{-1} + (1-p^{-3\nu_2})^{-1} - 1 \right) =$$

$$\prod_p (1-p^{2-3\nu_1-3\nu_2})^{-1} \cdot (1-p^{1-3\nu_1})^{-1} \cdot (1-p^{-3\nu_2})^{-1} \cdot (1-p^{1-3\nu_1-3\nu_2}) =$$

$$\zeta(3\nu_1+3\nu_2-2)\zeta(3\nu_1-1)\zeta(3\nu_2)\zeta(3\nu_1+3\nu_2-1)^{-1} .$$

Hence the contribution of the terms in question is:

$$\zeta(3\nu_1-1)\zeta(3\nu_2)\zeta(3\nu_1+3\nu_2-2) \cdot W_{0\,,\,0}^{(\nu_1,\nu_2)}(\tau, w_4)$$

Similarly, the contribution of the terms where $A_1 = 0$, $B_1 \neq 0$, $A_2 \neq 0$ is:

$$\zeta(3\nu_1)\zeta(3\nu_2-1)\zeta(3\nu_1+3\nu_2-2)\cdot W_{0,0}^{(\nu_1,\nu_2)}(\tau,w_5)$$

Adding the contributions of all terms, we obtain (7.6).

We now prove (7.7). First we consider the contribution of the terms $A_1 \neq 0$, $A_2 \neq 0$ to $\phi_{n_1,0}$. By a calculation similar to the nondegenerate case, this equals:

$$\zeta(3\nu_1+3\nu_2-1)\sum_{A_1=1}^{\infty}\sum_{A_2=1}^{\infty}\sum_{\substack{B_1,C_1 \bmod A_1 \\ B_2,C_2 \bmod A_2 \\ A_1C_2+B_1B_2+C_1A_2=0}} e\left(\frac{-n_1B_2}{A_2}\right)\cdot A_1^{-3\nu_1}A_2^{-3\nu_2}W_{n_1,0}^{(\nu_1,\nu_2)}(\tau,w_1) =$$

$$\zeta(3\nu_1+3\nu_2-1)\sum_{A_1=1}^{\infty}\sum_{A_2=1}^{\infty} r_{A_1,A_2}(A_1,-n_1)\cdot A_1^{-3\nu_1}A_2^{-3\nu_2}\cdot W_{n_1,0}^{(\nu_1,\nu_2)}(\tau,w_1) .$$

The sign here of n_1 is irrelevant, as $r_{A_1,A_2}(n_1,n_2) = r_{A_1,A_2}(|n_1|,|n_2|)$. Let $|n_1| = \prod_p p^{e(p)}$. We have, by (7.13):

$$\sum_{A_1=1}^{\infty}\sum_{A_2=1}^{\infty} r_{A_1,A_2}(A_1,-n_1)\cdot A_1^{-3\nu_1}A_2^{-3\nu_2} =$$

$$\prod_p \sum_{k_1=0}^{\infty}\sum_{k_2=0}^{\infty} r_{p^{k_1},p^{k_2}}(p^{k_1},p^{e(p)})\cdot p^{-3k_1\nu_1-3k_2\nu_2} =$$

$$\prod_p (1-p^{1-3\nu_1-3\nu_2})\cdot \sum_{k_1=0}^{\infty}\sum_{k_2=0}^{\infty} s_{p^{k_1},p^{k_2}}(p^{k_1},p^{e(p)})\cdot p^{-3k_1\nu_1-3k_2\nu_2}$$

We have:

$$s_{p^{k_1},p^{k_2}}(p^{k_1},p^{e(p)}) = C_{e(p)}(k_1,k_2)\cdot p^{k_1+k_2}$$

where:

$$C_e(k_1,k_2) = \begin{cases} k_1 + \min(k_2,e) + 1 - \max(k_1,k_2) & \text{if } k_1+e \geq k_2 \,; \\ 0 & \text{if } k_1+e < k_2 \,. \end{cases}$$

For convenience, we extend this definition so that $C_e(k_1,k_2) = 0$ if $k_2 < 0$. It is easy to see that:

$$C_e(k_1,k_2) = C_{e-1}(k_1,k_2-1) + C_0(k_1,k_2)$$

if $e > 0$. Thus,

$$C_e(k_1,k_2) = C_0(k_1,k_2) + C_0(k_1,k_2-1) + \ldots + C_0(k_1,k_2-e) \,.$$

Therefore:

$$\sum_{k_1=0}^{\infty}\sum_{k_2=0}^{\infty} s_{p^{k_1},p^{k_2}}(p^{k_1},p^{e(p)})\cdot p^{-3k_1\nu_1-3k_2\nu_2} =$$

$$(1+p^{1-3\nu_2}+p^{2-6\nu_2}+\ldots+p^{e(p)\cdot(1-3\nu_2)})\cdot$$

$$\sum_{k_1=0}^{\infty}\sum_{k_2=0}^{\infty} C_0(k_1,k_2)\cdot p^{k_1(1-3\nu_1)+k_2(1-3\nu_2)} =$$

$$\sigma_{1-3\nu_2}(p^{e(p)})\cdot \sum_{k_1=0}^{\infty}\sum_{k_2=0}^{k_1} p^{k_1(1-3\nu_1)+k_2(1-3\nu_2)} =$$

$$\sigma_{1-3\nu_2}(p^{e(p)})\cdot(1-p^{1-3\nu_1})^{-1}\cdot(1-p^{2-3\nu_1-3\nu_2})^{-1} .$$

Taking the product over all p, we see that the contribution of the terms in question is:

$$\zeta(3\nu_1)\zeta(3\nu_1-3\nu_2-2)\cdot\sigma_{1-3\nu_2}(|n_1|)\cdot W_{n_1,\,0}^{(\nu_1,\nu_2)}(\tau,w_1)$$

Next we calculate the contribution of the terms where $A_1 = 0$, $B_1,A_2 \neq 0$. Similar to the previous calculations, this equals:

$$\zeta(3\nu_1+3\nu_2-1)\sum_{B_1=1}^{\infty}\sum_{A_2=1}^{\infty}\sum_{\substack{C_1 \bmod B_1\\ B_2,C_2 \bmod A_2\\ B_1B_2+C_1A_2=0}} e(C_1n_1/B_1)\cdot B_1^{-3\nu_1}A_2^{-3\nu_2}\cdot W_{n_1,0}^{(\nu_1,\nu_2)}(\tau,w_5) .$$

For fixed B_1, A_2, we have:

$$\sum_{\substack{C_1 \bmod B_1 \\ B_2, C_2 \bmod A_2 \\ B_1B_2+C_1A_2=0}} e(C_1n_1/B_1) = \begin{cases} A_2\cdot(B_1,A_2) & \text{if } (B_1,A_2)\mid n_1 ; \\ 0 & \text{otherwise.} \end{cases}$$

Thus:

$$\sum_{B_1=1}^{\infty} \sum_{A_2=1}^{\infty} \sum_{\substack{C_1 \bmod B_1 \\ B_2, C_2 \bmod A_2 \\ B_1B_2+C_1A_2=0}} e(C_1n_1/B_1)\cdot B_1^{-3\nu_1} A_2^{-3\nu_2} =$$

$$\sum_{\substack{B_1, A_2=1 \\ (B_1,A_2)\mid n_1}}^{\infty} (B_1,A_2)\cdot B_1^{-3\nu_1} A_2^{1-3\nu_2} .$$

Writing $B_1 = b_1d$, $A_2 = a_2d$, where $d = (B_1,A_2)$, the last sum equals:

$$\sum_{d\mid n_1} d^{2-3\nu_1-3\nu_2}\cdot \sum_{\substack{b_1,a_2=1 \\ (b_1,a_2)=1}}^{\infty} b_1^{-3\nu_1} a_2^{1-3\nu_2} =$$

$$\sigma_{2-3\nu_1-3\nu_2}(|n_1|)\cdot \prod_p \Big(\sum_{k_1=0}^{\infty} p^{-k_1\nu_1} + \sum_{k_2=0}^{\infty} p^{k_2(1-3\nu_2)} - 1 \Big) =$$

$$\sigma_{2-3\nu_1-3\nu_2}(|n_1|)\cdot \zeta(3\nu_1)\zeta(3\nu_2-1)\zeta(3\nu_1+3\nu_2-1)^{-1} .$$

Hence the contribution of the terms in question is:

$$\zeta(3\nu_1)\zeta(3\nu_2-1)\sigma_{2-3\nu_1-3\nu_2}(|n_1|)\cdot W_{n_1,\,0}^{(\nu_1,\nu_2)}(\tau,w_5)\ .$$

Next we calculate the contribution of the terms $A_1 = A_2 = B_2 = 0$, $B_1 \neq 0$, $C_2 \neq 0$. Similar to previous calculations, this equals:

$$\zeta(3\nu_1+3\nu_2-1)\sum_{B_1=1}^{\infty}\sum_{C_2=1}^{\infty}\sum_{C_1 \bmod B_1} e(C_1n_1/B_1)\cdot B_1^{-3\nu_1}C_2^{-3\nu_2}\cdot W_{n_1,\,0}^{(\nu_1,\nu_2)}(\tau,w_3)\ .$$

Here:

$$\sum_{B_1=1}^{\infty}\sum_{C_2=1}^{\infty}\sum_{C_1 \bmod B_1} e(C_1n_1/B_1)\cdot B_1^{-3\nu_1}C_2^{-3\nu_2} =$$

$$\sum_{\substack{B_1=1\\ B_1|n_1}}^{\infty}\sum_{C_2=1}^{\infty} B_1^{1-3\nu_1}C_2^{-3\nu_2} = \sigma_{1-3\nu_1}(|n_1|)\zeta(3\nu_2)\ .$$

The contribution of the terms in question equals:

$$\zeta(3\nu_2)\zeta(3\nu_1+3\nu_2-1)\sigma_{1-3\nu_1}(|n_1|)\cdot W_{n_1,\,0}^{(\nu_1,\nu_2)}(\tau,w_3)\ .$$

We leave it to the reader to show that the remaining terms contribute zero to $\phi_{n_1,0}$. Adding the contributions of all terms, and applying (3.17), (3.20) and (3.22), we obtain (7.7).

The proof of (7.8) is similar.

It remains for us to prove Theorem 2.1, the analytic continuation and functional equations of the Eisenstein series. Let us first note that each Fourier term ϕ_{n_1,n_2} separately has meromorphic continuation and is invariant under the action (2.5) of the Weyl group.

The continuation and invariance of $\phi_{0,0}$ follow from (7.6) and from the functional equations (3.30-35). The continuation and invariance of $\phi_{n_1,0}(n_1 \neq 0)$ follow from (7.7), (3.36-39), together with the elementary fact that:

$$\sigma_\nu(n) = |n|^\nu \sigma_{-\nu}(n) .$$

Note that it is sufficient to verify invariance with respect to w_2 and w_3 since these elements generate the Weyl group. The continuation and invariance of ϕ_{0,n_2} are handled similarly.

Finally, we have the problem of the analytic continuation and invariance of ϕ_{n_1,n_2}, where $n_1,n_2 \neq 0$. From (3.29), it is sufficient to show that each coefficient a_{n_1,n_2} is invariant with respect to the Weyl group, and, by (7.10), one may assume that n_1 and n_2 are powers of the same prime p. In this case, the invariance follows from (7.9), together with the fact that the Weyl group simply permutes α, β, and γ.

Thus each Fourier term has meromorphic continuation to all ν_1,ν_2, and is invariant with respect to the Weyl group. To prove Theorem 7.1, however, we must actually show that the Fourier series (4.10) is convergent, and analyze more closely the polar divisor. In doing so, we shall obtain some results required by Bump and Goldfeld [2], relating the polar part of the GL(3) Eisenstein series, contributed by the degenerate terms in the Fourier expansion, to GL(2) Eisenstein series.

In (4.10), let:

$$H_{(\nu_1,\nu_2)}(\tau) = \sum_{g\in\Gamma_\infty^2\backslash\Gamma_1^2} \sum_{n_1=1}^{\infty} \sum_{n_2\neq 0} \phi_{n_1,n_2}(g\tau)$$

be the contribution of the nondegenerate terms. It follows from Theorem 2.1 that this series converges rapidly for all values of ν_1 and ν_2. We leave it for the reader to verify that this is indeed the case, but compare, for example, the second exercise following that theorem.

Let:

$$P_{(\nu_1,\nu_2)}(\tau) = \sum_{n_1=-\infty}^{\infty} \phi_{0,n_2}(\tau) + \sum_{g\in\Gamma_\infty^2\backslash\Gamma_1^2} \sum_{n_1=1}^{\infty} \phi_{n_1,0}(g\tau)$$

be the sum of all remaining terms, so that:

$$G_{(\nu_1,\nu_2)}(\tau) = H_{(\nu_1,\nu_2)}(\tau) + P_{(\nu_1,\nu_2)}(\tau)$$

We will analyze $P_{(\nu_1,\nu_2)}(\tau)$. It is convenient to introduce two complex variables as follows:

$$\tau_1 = x_1 + iy_1$$
$$\tau_2 = x_2 + iy_2 .$$

Also, it is convenient to introduce a modified divisor sum:

$$\tau_\nu(n) = |n|^{-\nu}\sigma_{2\nu}(|n|) = \tau_{-\nu}(n)$$

Let:

$$\lambda(\nu) = \pi^{-\nu/2}\Gamma(\tfrac{\nu}{2})\zeta(\nu) = \lambda(1-\nu)$$

by the functional equation of the Riemann zeta function. Now, if $\tau_0 = x + iy$ lies in the ordinary upper half-plane (homogeneous space of $SL(2, \mathbb{R})$), let:

$$G_\nu(\tau_0) = \tfrac{1}{2}\pi^{-\nu}\Gamma(\nu) \sum \frac{y^\nu}{|m\tau_0+n|^{2\nu}}$$

be the GL(2) Eisenstein series introduced in Chapter I. The Fourier expansion (1.12) may be rewritten:

$$G_\nu(\tau_0) = \lambda(2\nu)y^\nu + \lambda(2-2\nu)y^{1-\nu} + 2\sum_{\substack{n=-\infty \\ n\neq 0}}^{\infty} \tau_{\nu-\frac{1}{2}}(n)\sqrt{y}K_{\nu-\frac{1}{2}}(2\pi|n|y)e(nx) \tag{7.14}$$

Let:

$$\hat{G}_\nu(\tau_0) = 2\sum_{\substack{n=-\infty \\ n\neq 0}}^{\infty} \tau_{\nu-\frac{1}{2}}(n)\sqrt{y}K_{\nu-\frac{1}{2}}(2\pi|n|y)e(nx) .$$

This function is analytic for all values of ν, and is of rapid decay as $y \longrightarrow \infty$.

Let:

$$Q_{(\nu_1,\nu_2)}(\tau) = \tfrac{1}{2}\lambda(3\nu_2)\lambda(3\nu_1+3\nu_2-1)(y_1y_2^2)^{\frac{1}{2}(\nu_1+2\nu_2)} \cdot \sum_{(c,d)=1} |c\tau_2+d|^{-\frac{3}{2}(\nu_1+2\nu_2)} G_{\frac{3\nu_1}{2}}(cx_3+dx_1+i|c\tau_2+d|y_1) ,$$

the sum being over relatively prime pairs of integers c,d. This series is convergent if $re(\nu_2), re(\nu_1+\nu_2-\frac{1}{3}) > \frac{2}{3}$. However, this function has mermorphic continuation to all values of ν_1,ν_2, which we may prove as follows. Let:

$$\hat{Q}_{(\nu_1,\nu_2)}(\tau) = \tfrac{1}{2}\lambda(3\nu_2)\lambda(3\nu_1+3\nu_2-1)(y_1y_2^2)^{\frac{1}{2}(\nu_1+2\nu_2)} \sum_{(c,d)=1} |c\tau_2+d|^{-\frac{3}{2}(\nu_1+2\nu_2)} \hat{G}_{\frac{3\nu_1}{2}}(cx_3+dx_1+i|c\tau_2+d|y_1) .$$

This series is rapidly convergent for all values of ν_1,ν_2, although the zeta factors have poles. We find that:

$$Q_{(\nu_1,\nu_2)}(\tau) = \hat{Q}_{(\nu_1,\nu_2)}(\tau) + R_{(\nu_2,1-\nu_1-\nu_2)}(\tau) + R_{(1-\nu_1-\nu_2,\nu_1)}(\tau) \tag{7.15}$$

where:

$$R_{(\nu_1,\nu_2)}(\tau) = \lambda(3\nu_2-1)\lambda(3\nu_1+3\nu_2-2)(y_1^2y_2)^{\frac{1}{2}(2-\nu_1-2\nu_2)} G_{\frac{3\nu_1}{2}}(\tau_2) .$$

This proves the meromorphic continuation of $Q_{(\nu_1,\nu_2)}$. We will show that:

$$P_{(\nu_1,\nu_2)}(\tau) =$$

$$\hat{Q}_{(\nu_1,\nu_2)}(\tau) + \hat{Q}_{(\nu_2,1-\nu_1-\nu_2)}(\tau) + \hat{Q}_{(1-\nu_1-\nu_2,\nu_1)}(\tau)$$

$$+R_{(\nu_1,\nu_2)}(\tau) + R_{(\nu_2,1-\nu_1-\nu_2)}(\tau) + R_{(1-\nu_1-\nu_2,\nu_1)}(\tau) \tag{7.16}$$

By (7.15), this implies that:

$$P_{(\nu_1,\nu_2)}(\tau) =$$

$$Q_{(\nu_1,\nu_2)}(\tau) + Q_{(\nu_2,1-\nu_1-\nu_2)}(\tau) + Q_{(1-\nu_1-\nu_2,\nu_1)}(\tau)$$

$$-R_{(\nu_1,\nu_2)}(\tau) - R_{(\nu_2,1-\nu_1-\nu_2)}(\tau) - R_{(1-\nu_1-\nu_2,\nu_1)}(\tau) . \tag{7.17}$$

Let us prove (7.16). First note that since:

$$\begin{pmatrix} -1 & & \\ & -1 & \\ & & 1 \end{pmatrix} \in \Gamma_1^2$$

we may write:

$$\sum_{\Gamma_\infty^2 \backslash \Gamma_1^2} \sum_{n_1=1}^{\infty} \phi_{n_1,0}(g\cdot\tau) = \tfrac{1}{2} \sum_{\Gamma_\infty^2 \backslash \Gamma_1^2} \sum_{\substack{n_1=-\infty \\ n_1\neq 0}}^{\infty} \phi_{n_1,0}(g\cdot\tau)$$

By (7.7) and (3.40) this equals:

$$\tfrac{1}{2} \sum_{\Gamma_\infty^2 \backslash \Gamma_1^2} S_{(\nu_1,\nu_2)}(g\tau) + \tfrac{1}{2} \sum_{\Gamma_\infty^2 \backslash \Gamma_1^2} S_{(\nu_2,1-\nu_1-\nu_2)}(g\tau)$$
$$+ \tfrac{1}{2} \sum_{\Gamma_\infty^2 \backslash \Gamma_1^2} S_{(1-\nu_1-\nu_2,\nu_1)}(g\tau) \qquad (7.18)$$

where:

$$S_{(\nu_1,\nu_2)}(\tau) = 2\lambda(3\nu_2)\lambda(3\nu_1+3\nu_2-1)(y_1y_2^2)^{\frac{1}{2}(\nu_1+2\nu_2)}$$
$$\sum_{\substack{n_1=-\infty \\ n_1\neq 0}}^{\infty} \tau_{\frac{1}{2}(3\nu_1-1)}(n_1)\sqrt{y_1}K_{\frac{1}{2}(3\nu_1-1)}(2\pi|n_1|y_1)e(n_1x_1) \ .$$

By (7.14):

$$S_{(\nu_1,\nu_2)}(\tau) = \lambda(3\nu_1)\lambda(3\nu_1+3\nu_2-1)(y_1y_2^2)^{\frac{1}{2}(\nu_1+2\nu_2)}\hat{G}_{\frac{3\nu_1}{2}}(\tau_1) .$$

We now require specific knowledge of the action of Γ_1^2 on τ. We find that:

$$\begin{pmatrix} a & b & \\ c & d & \\ & & 1 \end{pmatrix}\begin{pmatrix} y_1y_1 & y_1x_2 & x_3 \\ & y_1 & x_1 \\ & & 1 \end{pmatrix} = \begin{pmatrix} y_1'y_2' & y_1'x_2' & x_2' \\ & y_1' & x_1' \\ & & 1 \end{pmatrix}k$$

where k is the orthogonal matrix:

$$\begin{pmatrix} (cx_2+d)\Delta^{-1} & -cy_2\Delta^{-1} & \\ cy_2\Delta^{-1} & (cx_2+d)\Delta^{-1} & \\ & & 1 \end{pmatrix} \qquad (\Delta = |c\tau_2+d|)$$

and:

$$x_2' + iy_2' = \frac{a\,\tau_2+b}{c\,\tau_2+d} \qquad x_1' = cx_3 + dx_1$$

$$y_1' = |c\,\tau_2+d|\,y_1 \qquad x_3' = ax_3 + bx_1$$

Thus by (7.18), we have:

$$\sum_{\Gamma_\infty^2 \backslash \Gamma_1^2} \sum_{n_1=1}^{\infty} \phi_{n_1,0}(g_1\tau) = \hat{Q}_{(\nu_1,\nu_2)}(\tau)+\hat{Q}_{(\nu_2,1-\nu_1-\nu_2)}(\tau)+Q_{(1-\nu_1-\nu_2,\nu_1)}(\tau) \tag{7.19}$$

On the other hand, by (7.8) and (3.45) we have:

$$\sum_{\substack{n_2=-\infty \\ n_2\neq 0}}^{\infty} \phi_{0,n_2}(\tau) = T_{(\nu_1,\nu_2)}(\tau) + T_{(\nu_2,1-\nu_1-\nu_2)}(\tau) + T_{(1-\nu_1-\nu_2,\nu_1)}(\tau)$$

where:

$$T_{(\nu_1,\nu_2)}(\tau) = 2\lambda(3\nu_1)\lambda(3\nu_1+3\nu_2-1)(y_1^2y_2)^{\frac{1}{2}(2\nu_1+\nu_2)}$$

$$\sum_{\substack{n_2=-\infty \\ n_2\neq 0}}^{\infty} \tau_{\frac{1}{2}(3\nu_2-1)}(n_1)K_{\frac{1}{2}(3\nu_2-1)}(2\pi|n_2|y_2)e(n_2x_2) .$$

And by (7.6), (3.30-35) and (3.10), we have:

$$\phi_{0,0}(\tau) = U_{(\nu_1,\nu_2)}(\tau) + U_{(\nu_2,1-\nu_1-\nu_2)}(\tau) + U_{(1-\nu_1-\nu_2,\nu_1)}(\tau)$$

where:

$$U_{(\nu_1,\nu_2)}(\tau) =$$

$$\lambda(3\nu_1)\lambda(3\nu_1+3\nu_2-1)(y_1^2y_2)^{\frac{1}{2}(2\nu_1+\nu_2)}[\lambda(3\nu_2)y_2^{\frac{3\nu_2}{2}} + (2-3\nu_2)y_2^{\frac{2-3\nu_2}{2}}] .$$

By (7.14) we have:

$$T_{(\nu_1,\nu_2)}(\tau) + U_{(\nu_1,\nu_2)}(\tau) = R_{(\nu_1,\nu_2)}(\tau)$$

Thus:

$$\sum_{n_2=-\infty}^{\infty} \phi_{0,n_2}(\tau) = R_{(\nu_1,\nu_2)}(\tau) + R_{(\nu_2,1-\nu_1-\nu_2)}(\tau) + R_{(1-\nu_1-\nu_2,\nu_1)}(\tau) .$$

Combining this with (7.19), we obtain (7.16).

Now, we may prove that the polar divisor of $G_{(\nu_1,\nu_2)}$ is as described in Theorem 2.1. We have noted that $G_{(\nu_1,\nu_2)}$ differs from $P_{(\nu_1,\nu_2)}$ by a holomorphic function. Referring to the definitions of $Q_{(\nu_1,\nu_2)}$ and $R_{(\nu_1,\nu_2)}$, the poles of $P_{(\nu_1,\nu_2)}$ may be read off from the poles of $\lambda(s)$, at $s = 0$ and 1. We see that there are six polar lines as described in the theorem. It is not a priori clear that $P_{(\nu_1,\nu_2)}$ does not also have three polar lines at $\nu_1,\nu_2,1-\nu_1-\nu_2 = \frac{1}{3}$; the six components of $P_{(\nu_1,\nu_2)}$ given in (7.16) also have polar divisors along those lines, which actually cancel in pairs. That cancellation occurs may be seen without any calculation as follows. $G_{(\nu_1,\nu_2)}$ is invariant with respect to the involution w_1, which fixes the line $1-\nu_1-\nu_2 = \frac{1}{3}$. Thus, the multiplicity of that line in the polar divisor must be even. On the other hand, (7.16) shows that this multiplicity is at most one, hence zero. Similarly, the other two lines each occur with zero multiplicity.

Let W be the group of substitutions in ν_1,ν_2 given by (2.5). W is normalized by the substitution $(\nu_1,\nu_2) \longrightarrow (\nu_2,\nu_1)$. Together, W and this substitution generate a group $\tilde{W}$ of order 12, namely, the complete group of automorphisms of the hexagon comprised of the six lines in Theorem 2.1. Let us note that $G_{(\nu_1,\nu_2)}$ actually satisfies functional equations for the extended group $\tilde{W}$, because:

$$G_{(\nu_2,\nu_1)}(\tau) = G_{(\nu_1,\nu_2)}({}^{\iota}\tau) . \tag{7.20}$$

This is an elementary consequence of (3.53).

It should be pointed out, however, that $H_{(\nu_1,\nu_2)}$ and $P_{(\nu_1,\nu_2)}$ satisfy no relation such as (7.20). This is because the Fourier expansion (4.10) is not stable under the involution ι.

For more information on the polar divisor of $G_{(\nu_1,\nu_2)}$, we refer the reader to the paper of Bump and Goldfeld [2].

CHAPTER VIII

THE ANALYTIC CONTINUATION AND FUNCTIONAL EQUATIONS SATISFIED BY THE L-SERIES ASSOCIATED WITH AN AUTOMORPHIC FORM

Let ϕ be an automorphic form of type (ν_1,ν_2), and let a_{n_1,n_2} be the matrix of Fourier coefficients, as in Chapter IV. We will consider the L-series:

$$L(s,\phi) = \sum_{n=1}^{\infty} a_{1,n} n^{-s} . \tag{8.1}$$

Also, let $\tilde{\phi}$ be the dual automorphic form defined by (4.14). By (4.15), we have:

$$L(s,\tilde{\phi}) = \sum_{n=1}^{\infty} a_{n,1} n^{-s} . \tag{8.2}$$

Let us define the following gamma factors:

$$\Phi(s) = \pi^{-\frac{3s}{2}} \Gamma\left(\frac{s+1-2\nu_1-\nu_2}{2}\right)\Gamma\left(\frac{s+\nu_1-\nu_2}{2}\right)\Gamma\left(\frac{s-1+\nu_1+2\nu_2}{2}\right)$$

$$\tilde{\Phi}(s) = \pi^{-\frac{3s}{2}} \Gamma\left(\frac{s+1-\nu_1-2\nu_2}{2}\right)\Gamma\left(\frac{s-\nu_1+\nu_2}{2}\right)\Gamma\left(\frac{s-1+2\nu_1+\nu_2}{2}\right) .$$

Our objective will be to prove:

THEOREM: The series (8.1) is convergent for large values of s. If ϕ is a cusp form, then $L(s,\phi)$ has analytic continuation to all values of s, and satisfies the functional equation:

$$\Phi(s)L(s,\phi) = \tilde{\Phi}(1-s)L(1-s,\tilde{\phi}) \tag{8.3}$$

If ϕ is not a cusp form, then $L(s,\phi)$ still has meromorphic continuation, and satisfies the same functional equation.

The functional equations were first proved by Godement and Jacquet [5]. However, their proof was somewhat indirect. A more satisfactory proof was obtained by Jacquet, Piatetski-Shapiro and Shalika [15]. It is the latter proof which we shall follow (except that their automorphic forms are on the adele group).

First we need to know something about the growth of the Fourier coefficients a_{n_1,n_2}. If ϕ is a cusp form, we shall prove that:

$$a_{n_1,n_2} = O(|n_1 n_2|) \tag{8.4}$$

Remark: Even if ϕ is not cuspidal, the Fourier coefficients have at most polynomial growth in n_1,n_2, which is all that is required for the convergence (8.1) with s sufficiently large.

To prove (8.4), let us first note that, being cuspidal, ϕ is bounded on $\mathcal{H}$. Indeed, if c is sufficiently small, the so-called "Siegel set":

$$\{\tau \mid y_1,y_2 > c, -\tfrac{1}{2} \leq x_1,x_2,x_3 \leq \tfrac{1}{2}\}$$

contains a fundamental domain for Γ. Given the vanishing of the degenerate terms of the Fourier expansion, the boundedness of ϕ on this set follows from (4.10) and the estimate of Theorem 2.1 for the Whittaker function. Given that ϕ is bounded on $\mathcal{H}$, we have:

$$|n_1 n_2|^{-1} a_{n_1,n_2} W_{1,1}^{(\nu_1,\nu_2)}\begin{pmatrix}1 & & \\ & 1 & \\ & & 1\end{pmatrix} =$$

$$\int_0^1\int_0^1\int_0^1 \phi\left(\begin{pmatrix}1 & \xi_2 & \xi_3 \\ & 1 & \xi_1 \\ & & 1\end{pmatrix}\begin{pmatrix}y_1y_2n_1^{-1}n_2^{-1} & & \\ & y_1n_1^{-1} & \\ & & 1\end{pmatrix}\right) e(-n_1\xi_1-n_2\xi_2)\,d\xi_1 d\xi_2 d\xi_3$$

$$= O(1)$$

whence (8.4).

It follows from (8.4) that the series (8.1) is convergent, at least for large values of s.

Let us use freely the notations of Chapter IV. Our first object is to prove that:

$$\int_0^1\int_0^1 \phi\left(\begin{pmatrix}1 & & \\ x & 1 & x_1 \\ & & 1\end{pmatrix}\tau\right) e(-x_1)\,dx\,dx_1 =$$

$$\int_{-\infty}^{\infty} \phi_1^0\left(\begin{pmatrix}1 & & \\ x & 1 & \\ & & 1\end{pmatrix}\tau\right) dx . \qquad (8.5)$$

Indeed, by (4.5), the right-hand side equals:

$$\sum_{m=-\infty}^{\infty}\int_0^1 \phi_1^0\left(\begin{pmatrix}1 & & \\ m & 1 & \\ & & 1\end{pmatrix}\begin{pmatrix}1 & & \\ x & 1 & \\ & & 1\end{pmatrix}\tau\right) dx =$$

$$\int_0^1 \sum_{m=-\infty}^{\infty} \phi_1^m\left(\begin{pmatrix}1 & & \\ x & 1 & \\ & & 1\end{pmatrix}\tau\right) dx .$$

By (4.1) and (4.3), we have:

$$\sum_{m=-\infty}^{\infty} \phi_1^m(\tau) = \int_0^1 \phi\left(\begin{pmatrix}1 & & \\ & 1 & x_1 \\ & & 1\end{pmatrix}\tau\right) e(-x_1)dx_1 .$$

(8.5) now follows.

Now, let us show that:

$$\phi_1^0\begin{pmatrix}t & & \\ & 1 & \\ & & 1\end{pmatrix} = \int_{-\infty}^{\infty} \tilde{\phi}_1^0\left(\begin{pmatrix}1 & & \\ x & 1 & \\ & & 1\end{pmatrix}\begin{pmatrix}t^{-1} & & \\ & 1 & \\ & & 1\end{pmatrix}\right) dx . \qquad (8.6)$$

Indeed, we have:

$$\phi\begin{pmatrix}t & & x_3 \\ & 1 & x_1 \\ & & 1\end{pmatrix} = \tilde{\phi}\begin{pmatrix}1 & -x_1 & -t^{-1}x_3 \\ & 1 & \\ & & t^{-1}\end{pmatrix} .$$

Now, since $\tilde{\phi}$ is left-invariant by Γ and right-invariant by K, this equals:

$$\tilde{\phi}\left(\begin{pmatrix} & & -1 \\ & 1 & \\ -1 & & \end{pmatrix}\begin{pmatrix} 1 & -x_1 & -t^{-1}x_3 \\ & 1 & \\ & & t^{-1} \end{pmatrix}\begin{pmatrix} & 1 & \\ & & 1 \\ -1 & & \end{pmatrix}\right) =$$

$$\tilde{\phi}\left(\begin{pmatrix} 1 & & \\ x_3 & 1 & x_1 \\ & & 1 \end{pmatrix}\begin{pmatrix} t^{-1} & & \\ & 1 & \\ & & 1 \end{pmatrix}\right).$$

Thus, by (8.5), we have:

$$\phi_1^0\begin{pmatrix} t & & \\ & 1 & \\ & & 1 \end{pmatrix} = \int_0^1\int_0^1 \phi\begin{pmatrix} t & & x_3 \\ & 1 & x_1 \\ & & 1 \end{pmatrix} e(-x_1)dx_1dx_3 =$$

$$\int_0^1\int_0^1 \tilde{\phi}\left(\begin{pmatrix} 1 & & \\ x_3 & 1 & x_1 \\ & & 1 \end{pmatrix}\begin{pmatrix} t^{-1} & & \\ & 1 & \\ & & 1 \end{pmatrix}\right) e(-x_1)dx_1dx_3 =$$

$$\int_{-\infty}^{\infty} \tilde{\phi}_1^0\left(\begin{pmatrix} 1 & & \\ x & 1 & \\ & & 1 \end{pmatrix}\begin{pmatrix} t^{-1} & & \\ & 1 & \\ & & 1 \end{pmatrix}\right) dx$$

whence (8.6).

Now, following Jacquet, Piatetski-Shapiro and Shalika, let us introduce two auxiliary Mellin transforms:

$$\Psi(s) = \int_0^{\infty} W\begin{pmatrix} y & & \\ & 1 & \\ & & 1 \end{pmatrix} y^{s-1}\,\frac{dy}{y}$$

$$\tilde{\Psi}(s) = \int_0^{\infty}\int_{-\infty}^{\infty} \widetilde{W}\begin{pmatrix} t & & \\ x & 1 & \\ & & 1 \end{pmatrix} t^{s-1}dx\frac{dt}{t}$$

where we denote:

$$W(\tau) = W^{(\nu_1,\nu_2)}_{1,1}(\tau,w_1)$$

$$\tilde{W}(\tau) = W^{(\nu_2,\nu_1)}_{1,1}(\tau,w_1) .$$

It follows easily from Theorem 2.1 that $\psi(s)$ converges for sufficiently large s. The convergence of $\tilde{\psi}(s)$ merits proof. Let us compute the coordinates of the matrix:

$$\begin{pmatrix} t & & \\ x & 1 & \\ & & 1 \end{pmatrix}$$

By multiplying on the right by the orthogonal similitude:

$$\begin{pmatrix} 1 & x & \\ -x & 1 & \\ & & \Delta \end{pmatrix} \qquad (\Delta = \sqrt{x^2+1}).$$

We see that the coordinates are:

$$x_1 = x_3 = 0 ; \qquad x_2 = tx\Delta^{-2}$$

$$y_1 = \Delta$$

$$y_2 = t\Delta^{-2} .$$

Now, the convergence of the integral for $\tilde{\Psi}$ follows easily from Theorem 2.1 for sufficiently large s.

Remark: The evaluation of the Mellin transforms Ψ and $\tilde{\Psi}$ is an open problem. The corresponding p-adic problem was solved by Jacquet, Piatetski-Shapiro and Shalika. Given an admissible representation ρ of GL(3) over a p-adic field, they showed that the Mellin transforms corresponding to Ψ and $\tilde{\Psi}$ of the p-adic Whittaker functions associated with ρ and its contragedient $\tilde{\rho}$ actually equal the local zeta functions associated with ρ and $\tilde{\rho}$ respectively. In the archimedean case, one might expect, by analogy, that Ψ and $\tilde{\Psi}$ would also equal the local zeta functions associated with the admissible representations parametrized by (ν_1,ν_2) and (ν_2,ν_1) respectively, i.e., the gamma factors denoted Φ and $\tilde{\Phi}$. Unable to prove this, they nevertheless were able to prove the integral formula (8.10) below, which is sufficient for the functional equations. We no longer believe that Ψ and Φ are essentially equal, however. The reason for this lack of belief is that the inverse Mellin transform of Φ satisfies a certain third order differential equation. We have made explicit the differential equations in Chapter II above, and while:

$$W\begin{pmatrix} y & & \\ & 1 & \\ & & 1 \end{pmatrix}$$

satisfies a sixth order differential equation, we see no reason that it should satisfy a third order differential equation. We believe that the functions Φ and Ψ are actually very closely related, but they are not constant multiples of each other.

On the other hand, we will evaluate the double Mellin transform:

$$\int_0^\infty\int_0^\infty W\begin{pmatrix} y_1 y_2 & & \\ & y_1 & \\ & & 1 \end{pmatrix} y_1^{s_1-1} y_2^{s_2-1} \frac{dy_1}{y_1} \frac{dy_2}{y_2}$$

in Chapter X.

Now, we will prove that $\Psi(s)\cdot L(s,\phi)$ has analytic continuation to all values of s, and satisfies the following preliminary functional equation:

$$\Psi(s)\cdot L(s,\phi) = \tilde{\Psi}(1-s)L(1-s,\tilde{\phi}) \tag{8.7}$$

We will prove this by evaluating the Mellin transform:

$$\int_0^\infty \phi_1^0\begin{pmatrix} y & & \\ & 1 & \\ & & 1 \end{pmatrix} y^{s-1} \frac{dy}{y}$$

in two different ways. First let us show that the integrand:

$$\phi_1^0\begin{pmatrix} y & & \\ & 1 & \\ & & 1 \end{pmatrix}$$

is rapidly decreasing as $y \longrightarrow 0$ or ∞. By (4.13), we have:

$$\phi_1^0\begin{pmatrix} y & & \\ & 1 & \\ & & 1 \end{pmatrix} = \sum_{n=-\infty}^{\infty} \phi_{1,n}\begin{pmatrix} y & & \\ & 1 & \\ & & 1 \end{pmatrix} = 2\sum_{n=1}^{\infty} \phi_{1,n}\begin{pmatrix} y & & \\ & 1 & \\ & & 1 \end{pmatrix}$$

so:

$$\phi_1^0\begin{pmatrix} y & & \\ & 1 & \\ & & 1\end{pmatrix} = 2\sum_{n=1}^{\infty} a_{1,n}n^{-1}W\begin{pmatrix} ny & & \\ & 1 & \\ & & 1\end{pmatrix} \tag{8.8}$$

It follows from Theorem 2.1 that the integrand is rapidly decreasing as $y \longrightarrow \infty$. On the other hand, by (8.6) and (4.15) the integrand equals:

$$\int_{-\infty}^{\infty} \tilde{\phi}_1^0\begin{pmatrix} y^{-1} & & \\ xy^{-1} & 1 & \\ & & 1\end{pmatrix}dx = 2\sum_{n=1}^{\infty} a_{n,1}n^{-1}\int_{-\infty}^{\infty} \tilde{W}\begin{pmatrix} ny^{-1} & & \\ xy^{-1} & 1 & \\ & & 1\end{pmatrix}dx$$

so after a change of variables:

$$\phi_1^0\begin{pmatrix} y & & \\ & 1 & \\ & & 1\end{pmatrix} = 2y\sum_{n=1}^{\infty} a_{n,1}n^{-1}\int_{-\infty}^{\infty} \tilde{W}\begin{pmatrix} ny^{-1} & & \\ x & 1 & \\ & & 1\end{pmatrix}dx . \tag{8.9}$$

As in our proof of the convergence of the integral for $\tilde{\Psi}$ this is rapidly decreasing as $y \longrightarrow 0$.

Thus, the integral:

$$\int_0^{\infty} \phi_1^0\begin{pmatrix} y & & \\ & 1 & \\ & & 1\end{pmatrix} y^{s-1}\,\frac{dy}{y}$$

is convergent for all values of s. If s is large, we may evaluate this by means of (8.8); it equals:

$$\Psi(s)\cdot L(s,\phi)$$

And on the other hand, if $-s$ is large, we may evaluate the integral by means of (8.9); it equals:

$$\tilde{\Psi}(1-s)L(1-s,\tilde{\phi})$$

whence (8.7).

Now, (8.3) will follow if we prove that $\Psi(s)\tilde{\Phi}(1-s)$ has analytic continuation to all values of s, and:

$$\Psi(s)\tilde{\Phi}(1-s) = \Phi(s)\tilde{\Psi}(1-s) \tag{8.10}$$

This integral formula was obtained by Jacquet, Piatetski-Shapiro and Shalika. We will give a proof which is different from theirs, however.

Firstly, we will generalize our proof of (8.7) to the case where ϕ is the Eisenstein series:

$$\phi(\tau) = G_{(\nu_1,\nu_2)}(\tau)$$

Now, in that case, we will show, using the explicit formula of Theorem 7.2 for the Fourier coefficients of the Eisenstein series, that:

$$L(s,\phi) = \zeta(s+1-2\nu_1-\nu_2)\zeta(s+\nu_1-\nu_2)\zeta(s-1+\nu_1+2\nu_2) \tag{8.11}$$

Thus, (8.3) is valid for the Eisenstein series at least. Thus, we have both (8.3) and (8.7) for the Eisenstein series; consequently, we have (8.10).

Let us generalize our previous proof of (8.7) to the Eisenstein series. First, suppose s is large. We have:

$$2\Psi(s)L(s,\phi) = \int_0^\infty \{\phi_1^0\begin{pmatrix} y & & \\ & 1 & \\ & & 1\end{pmatrix} - \phi_{1,0}\begin{pmatrix} y & & \\ & 1 & \\ & & 1\end{pmatrix}\}\, y^{s-1}\,\frac{dy}{y}\,.$$

By (8.6), this equals:

$$\int_1^\infty \{\phi_1^0\begin{pmatrix} y & & \\ & 1 & \\ & & 1\end{pmatrix} - \phi_{1,0}\begin{pmatrix} y & & \\ & 1 & \\ & & 1\end{pmatrix}\}\, y^{s-1}\,\frac{dy}{y}$$

$$+\int_1^\infty\int_{-\infty}^\infty \{\tilde{\phi}_1^0\begin{pmatrix} y & & \\ x & 1 & \\ & & 1\end{pmatrix} - \tilde{\phi}_{1,0}\begin{pmatrix} y & & \\ x & 1 & \\ & & 1\end{pmatrix}\}\, y^{-s}dx\frac{dy}{y}$$

$$-\int_0^1 \phi_{1,0}\begin{pmatrix} y & & \\ & 1 & \\ & & 1\end{pmatrix} y^{s-1}\,\frac{dy}{y}$$

$$+\int_1^\infty\int_{-\infty}^\infty \tilde{\phi}_{1,0}\begin{pmatrix} y & & \\ x & 1 & \\ & & 1\end{pmatrix} y^{-s}dx\frac{dy}{y}$$

Now, we have, by (7.7) and (3.40):

$$\int_0^1 \phi_{1,0}\begin{pmatrix} y & & \\ & 1 & \\ & & 1\end{pmatrix} y^{s-1}\,\frac{dy}{y} = A_{(\nu_1,\nu_2)}(s)+A_{(\nu_2,1-\nu_1-\nu_2)}(s)+A_{(1-\nu_1-\nu_2,\nu_1)}(s)$$

where:

$$A_{(\nu_1,\nu_2)}(s) = 2\lambda(3\nu_2)\lambda(3\nu_1+3\nu_2-1)K_{\frac{3\nu_1-1}{2}}(2\pi)(\nu_1+2\nu_2-1+s)^{-1}$$

$$(\lambda(s) = \pi^{-\frac{s}{2}}\Gamma(\tfrac{s}{2})\zeta(s)) .$$

Also:

$$\int_1^\infty \int_{-\infty}^\infty \tilde{\phi}\begin{pmatrix} y & & \\ x & 1 & \\ & & 1 \end{pmatrix} y^{-s} dx \frac{dy}{y} =$$

$$B_{(\nu_1,\nu_2)}(s) + B_{(\nu_2,1-\nu_1-\nu_2)}(s) + B_{(1-\nu_1-\nu_2,\nu_1)}(s)$$

where:

$$B_{(\nu_1,\nu_2)}(s) =$$

$$2\lambda(3\nu_1)\lambda(3\nu_1+3\nu_2-1) \int_{-\infty}^\infty (\sqrt{x^2+1})^{\frac{1}{2}-3\nu_1-\frac{3\nu_2}{2}} K_{\frac{3\nu_2-1}{2}}(2\pi\sqrt{x^2+1})dx\cdot(s-\nu_1-2\nu_2)^{-1}$$

This may be deduced from (7.7) and (3.40) also. Thus:

$$2\Psi(s)L(s,\phi) =$$

$$\int_1^\infty \{\phi_1^0\begin{pmatrix} y & & \\ & 1 & \\ & & 1\end{pmatrix} - \phi_{1,0}\begin{pmatrix} y & & \\ & 1 & \\ & & 1\end{pmatrix}\}y^{s-1}\frac{dy}{y}$$

$$+\int_1^\infty\int_{-\infty}^\infty \{\tilde\phi_1^0\begin{pmatrix} y & & \\ x & 1 & \\ & & 1\end{pmatrix} - \tilde\phi_{1,0}\begin{pmatrix} y & & \\ x & 1 & \\ & & 1\end{pmatrix}\}y^{-s}dx\,\frac{dy}{y}$$

$$- A_{(\nu_1,\nu_2)}(s) - A_{(\nu_2,1-\nu_1-\nu_2)}(s) - A_{(1-\nu_1-\nu_2,\nu_1)}(s)$$

$$+ B_{(\nu_1,\nu_2)}(s) + B_{(\nu_2,1-\nu_1-\nu_2)}(s) + B_{(1-\nu_1-\nu_2,\nu_1)}(s) \,. \qquad (8.12)$$

From this follows the analytic continuation of $\Psi(s)\cdot L(s,\phi)$.

Similarly, if we assume that $-s$ is large:

$$2L(1-s,\tilde\phi)\tilde\Psi(1-s) =$$

$$\int_1^\infty \{\phi_1^0\begin{pmatrix} y & & \\ & 1 & \\ & & 1\end{pmatrix} - \phi_{1,0}\begin{pmatrix} y & & \\ & 1 & \\ & & 1\end{pmatrix}\}y^{s-1}\frac{dy}{y}$$

$$+\int_1^\infty\int_{-\infty}^\infty \{\tilde\phi_1^0\begin{pmatrix} y & & \\ x & 1 & \\ & & 1\end{pmatrix} - \tilde\phi_{1,0}\begin{pmatrix} y & & \\ x & 1 & \\ & & 1\end{pmatrix}\}y^{-s}dx\,\frac{dy}{y}$$

$$+\int_1^\infty \phi_{1,0}\begin{pmatrix} y & & \\ & 1 & \\ & & 1\end{pmatrix}y^{s-1}\frac{dy}{y} - \int_0^1\int_{-\infty}^\infty \tilde\phi_{1,0}\begin{pmatrix} y & & \\ x & 1 & \\ & & 1\end{pmatrix}y^{-s}dx\,\frac{dy}{y}$$

We have:

$$\int_1^\infty \phi_{1,0}\begin{pmatrix} y & & \\ & 1 & \\ & & 1\end{pmatrix}y^{s-1}\frac{dy}{y} = -A_{(\nu_1,\nu_2)}(s) - A_{(\nu_2,1-\nu_1-\nu_2)}(s) - A_{(1-\nu_1-\nu_2,\nu_1)}(s)$$

and:

$$\int_0^1\int_{-\infty}^{\infty} \tilde{\phi}_{1,0}\begin{pmatrix} y & & \\ x & 1 & \\ & & 1\end{pmatrix} y^{-s}dx\,\frac{dy}{y} = -B_{(\nu_1,\nu_2)}(s)-B_{(\nu_2,1-\nu_1-\nu_2)}(s)-B_{(1-\nu_1-\nu_2,\nu_1)}(s)$$

Thus, we have:

$$2\tilde{\Psi}(1-s)L(1-s,\tilde{\phi}) = \int_1^{\infty} \{\phi_1^0\begin{pmatrix} y & & \\ & 1 & \\ & & 1\end{pmatrix} - \phi_{1,0}\begin{pmatrix} y & & \\ & 1 & \\ & & 1\end{pmatrix} y^{s-1}\frac{dy}{y}$$

$$+\int_1^{\infty}\int_{-\infty}^{\infty} \{\tilde{\phi}_1^0\begin{pmatrix} y & & \\ x & 1 & \\ & & 1\end{pmatrix} - \tilde{\phi}_{1,0}\begin{pmatrix} y & & \\ x & 1 & \\ & & 1\end{pmatrix} y^{-s}dx\,\frac{dy}{y}$$

$$- A_{(\nu_1,\nu_2)}(s) - A_{(\nu_2,1-\nu_1-\nu_2)}(s) - A_{(1-\nu_1-\nu_2,\nu_1)}(s)$$

$$+ B_{(\nu_1,\nu_2)}(s) + B_{(\nu_2,1-\nu_1-\nu_2)}(s) + B_{(1-\nu_1-\nu_2,\nu_1)}(s)$$

This is identical with (8.12), whence (8.7).

(8.11) remains to be proved. By (7.9), we have:

$$L(s,\phi) = \prod_p \sum_{k=0}^{\infty} a_{1,p^k}\cdot p^{-ks} = \prod_p \sum_{k=0}^{\infty} S_{0,k}(p^{\alpha},p^{\beta},p^{\gamma})p^{-ks}$$

where:

$$\alpha = -\nu_1 - 2\nu_2 + 1$$

$$\beta = -\nu_1 + \nu_2$$

$$\gamma = 2\nu_1 + \nu_2 - 1\ .$$

By (6.5), we have:

$$S_{0,k}(a,b,c) = (b-a)^{-1}(c-b)^{-1}(a-c)^{-1}\{a^{2+k}(b-c)+b^{2+k}(c-a)+c^{2+k}(a-b)\}$$

Thus, summing the geometric series:

$$\sum_{k=0}^{\infty} S_{0,k}(a,b,c)x^k = (b-a)^{-1}(c-b)^{-1}(a-c)^{-1}.$$

$$[(1-ax)^{-1}a^2(b-c) + (1-bx)^{-1}b^2(c-a) + (1-cx)^{-1}c^2(a-b)] =$$

$$(1-ax)^{-1}(1-bx)^{-1}(1-cx)^{-1} .$$

Consequently, our previous infinite product equals:

$$\prod_p (1-p^{\alpha-s})^{-1}(1-p^{\beta-s})^{-1}(1-p^{\gamma-s})^{-1} =$$

$$\zeta(s-\alpha)\zeta(s-\beta)\zeta(s-\gamma)$$

as required.

This proves (8.11). (8.10) and (8.3) now follow.

CHAPTER IX

HECKE OPERATORS AND L-SERIES

In the last chapter, we proved the analytic continuation and functional equations of the L-series 8.1. The L-series have another important aspect, namely, the Euler products associated with a Hecke eigenform. We will see that a certain algebra of arithmetically defined operators, the Hecke algebra, acts on the cusp forms of type (ν_1,ν_2), and that the space of such cusp forms has a basis consisting of Hecke eigenforms. If ϕ is such an eigenform, normalized so that, if a_{n_1,n_2} is the matrix of Fourier coefficients, then the leading coefficient $a_{1,1} = 1$, then, we will show that:

$$L(s,\phi) = \prod_p (1-a_{1,p}p^{-s}+a_{p,1}p^{-2s}-p^{-3s})^{-1} \tag{9.1}$$

We will also consider the double L-series formed with the matrix of Fourier coefficients. We will prove the following formula:

$$\sum_{n_1=1}^{\infty}\sum_{n_2=1}^{\infty} a_{n_1,n_2} n_1^{-s_1} n_2^{-s_2} = \frac{L(s_1,\tilde{\phi})L(s_2,\phi)}{\zeta(s_1+s_2)} \tag{9.2}$$

Thus, all the coefficients a_{n_1,n_2} are determined by those coefficients occurring in $L(s,\phi)$ and $L(s,\tilde{\phi})$. Since, as we saw in the previous chapter, the second L-series is actually expressed in terms of the first by the functional equation, we may say that an automorphic form is determined by its L-series.

We will also prove, that if ϕ is a normalized Hecke eigenform, then the Fourier coefficients of ϕ are multiplicative:

$$\text{If} \quad (n_1n_2, n_1'n_2') = 1 \quad \text{then} \quad a_{n_1n_1', n_2n_2'} = a_{n_1,n_2}a_{n_1',n_2'} \tag{9.3}$$

Furthermore, for each rational p, if we factor the local factor of the L-series:

$$1 - a_{1,p}p^{-s} + a_{p,1}p^{-2s} - p^{-3s} = (1-\alpha p^{-s})(1-\beta p^{-s})(1-\gamma p^{-s}) \tag{9.4}$$

then, in terms of the Schur polynomials of Chapter VI, we have:

$$a_{p^{k_1},p^{k_2}} = S_{k_1,k_2}(\alpha,\beta,\gamma) \tag{9.5}$$

The Hecke operators on $GL(n)$ were first studied by Shimura and by Tamagawa (cf. [32]). A convenient reference for the algebraic theory is Shimura [30]. To avoid duplication of a standard and widely known text, we shall assume of the reader a thorough knowledge of the relevant portions of the latter work, especially section 3.2. We will somewhat modify Shimura's notation, however. Furthermore, we will leave routine verifications to the reader (though there is much that is not routine). With this caveat, our proofs will be complete.

Let $\Gamma = GL(3,\mathbb{Z})$, $\overline{\Gamma} = GL(3,\mathbb{Q})$, and let Δ be the semigroup of 3×3 integral nonsingular matrices. Let $H = R(\Gamma,\Delta)$ and $\overline{H} = R(\Gamma,\overline{\Gamma})$ be the corresponding $\mathbb{Z}$-algebras of double cosets (see [30] for definitions and proofs of the algebraic properties required). It is known that H and $\overline{H}$ are commutative rings. Let H_p (resp. $\overline{H}_p$) denote the subalgebra of H (resp. $\overline{H}_p$) corresponding to those double cosets whose elementary divisors are powers of a given prime p. We have $H = \bigotimes_p H_p$ and $\overline{H} = \bigotimes_p \overline{H}_p$ (internal tensor product of $\mathbb{Z}$-algebras). Furthermore, the structure of H_p and $\overline{H}_p$ is known. If n is a positive integer, let:

$$T_n = \sum_{n_0^3 n_1^2 n_2 = n} \Gamma\begin{pmatrix} n_0 n_1 n_2 & & \\ & n_0 n_1 & \\ & & n_0 \end{pmatrix}\Gamma$$

$$S_n = \sum_{n_0^3 n_1^2 n_2 = n} \Gamma\begin{pmatrix} n_0^2 n_1^2 n_2 & & \\ & n_0^2 n_1 n_2 & \\ & & n_0^2 n_1 \end{pmatrix}\Gamma$$

$$R_n = \Gamma\begin{pmatrix} n & & \\ & n & \\ & & n \end{pmatrix}\Gamma$$

The peculiar expression for S_n is explained by (9.7) below. We find that $H_p = \mathbb{Z}[T_p,S_p,R_p]$ is a polynomial ring in three variables; $\overline{H}_p$ is obtained from H_p by adjoining R_p^{-1}. Furthermore, we have the following formal identity (due to Tamagawa):

$$\sum_n T_n\cdot n^{-s} = \prod_p (1-T_p\cdot p^{-s}+S_p\cdot p^{1-2s}-R_p\cdot p^{3-3s})^{-1} \tag{9.6}$$

The involution ι of $\overline{\Gamma}$ (cf. (2.2)) induces an involutory automorphism of $\overline{H}$, with the effect:

$$\begin{aligned} T_n &\dashrightarrow R_n^{-1}\cdot S_n \\ S_n &\dashrightarrow R_n^{-1}\cdot T_n \\ R_n &\dashrightarrow R_n^{-1} \end{aligned} \tag{9.7}$$

Unfortunately, H is not stable under this automorphism. However, since we shall be concerned with a representation of H on a space of automorphic forms on which R_n acts trivially, this circumstance will be of little consequence to us.

The action of the Hecke operators on automorphic forms is defined in the usual way: Let ϕ be an automorphic form as defined in Chapter III. If $\Gamma\alpha\Gamma = \bigcup_i \Gamma\alpha_i$, then by definition:

$$(\phi|\Gamma\alpha\Gamma)(\tau) = \sum_i \phi(\alpha_i\tau) .$$

We will need to know the effect of the Hecke operators on the Fourier coefficients of an automorphic form. Let us fix an integer n, and let a_{n_1,n_2} be the matrix of Fourier coefficients of ϕ, b_{n_1,n_2} the matrix of Fourier coefficients of $\phi|T_n$, and c_{n_1,n_2} the matrix of Fourier coefficients of $\phi|S_n$. We will show that:

$$b_{n_1,n_2} = n \sum_{\substack{p_0p_1p_2=n \\ p_1|n_1 \\ p_2|n_2}} a_{\frac{n_1p_0}{p_1},\frac{n_2p_1}{p_2}} \tag{9.8}$$

and:

$$c_{n_1,n_2} = n \sum_{\substack{p_0p_1p_2=n \\ p_0|n_1 \\ p_1|n_2}} a_{\frac{n_1p_1}{p_0},\frac{n_2p_2}{p_1}} \tag{9.9}$$

So as not to disrupt the continuity of the exposition, we will defer the proofs of these formulae to the end of the chapter. Let us assume them for the present.

Although (9.8) and (9.9) are valid without restriction, their consequences are most dramatic in the case where ϕ is a Hecke eigenform. Let us recall, therefore, why eigenforms exist.

Now, there exists an inner product, known as the <u>Petersson inner product</u>, on the cusp forms, namely, if ϕ and ψ are cusp forms, let:

$$\langle \phi,\psi\rangle = \int_{\Gamma\backslash\mathcal{H}} \phi(\tau)\overline{\psi(\tau)}\, d\mu(\tau)$$

where:

$$d\mu(\tau) = y_1^{-3}y_2^{-3}dy_1dy_2dx_1dx_2dx_3$$

is the invariant metric on $\mathcal{H}$. For our purposes, we may assume that ϕ and ψ are forms of the same type (ν_1,ν_2) and indeed, if they are not, they will be orthogonal.

Now, with the involution ι acting on the Hecke algebra as before, we will prove that, if T is any Hecke operator, then ${}^{\iota}T$ is the adjoint of T with respect to the Petersson inner product. Indeed, we may assume that $T = \Gamma\alpha\Gamma = \bigcup_i \Gamma\alpha_i$ is a double coset. By the theory of elementary divisors, it is known that $\Gamma\alpha\Gamma = \Gamma{}^t\alpha\Gamma$, and furthermore, the representatives α_i may be chosen in such a way that $T = \bigcup_i \Gamma\alpha_i = \bigcup_i \Gamma{}^t\alpha_i$. Now, let Γ_1 be a subgroup of finite index such that:

$$\tau \dashrightarrow \phi({}^t\alpha_i w_1 \tau)\overline{\psi(\tau)}$$

is invariant under Γ_1 for each α_i. Then:

$$\begin{aligned}
&\langle \phi | T, \psi\rangle \\
&= \frac{1}{[\Gamma:\Gamma_1]} \sum_i \int_{\Gamma_1\backslash\mathcal{H}} \phi({}^t\alpha_i w_1 \tau)\overline{\psi(\tau)}\, d\mu(\tau) \\
&= \frac{1}{[\Gamma:\Gamma_1]} \sum_i \int_{\Gamma_1\backslash\mathcal{H}} \phi(\tau)\overline{\psi({}^t\alpha_i^{-1} w_1 \tau)}\, d\mu(\tau) \\
&= \langle \phi, \psi | {}^{\iota}T\rangle
\end{aligned}$$

Thus ${}^{\iota}T$ is indeed the adjoint of T with respect to the Petersson inner product. For example, S_n is the adjoint of T_n.

Now, the Hecke operators form a commutative algebra of normal operators, and may thus be simultaneously diagonalized. Furthermore, they also commute with the commutative algebra of differential operators introduced in Chapter II. It follows that the space of cusp forms of type (ν_1, ν_2) has a basis consisting of simultaneous eigenforms for all Hecke operators.

Let us assume now that ϕ is a normalized Hecke eigenform. We will explore the consequences of (9.8-9). Let us firstly note that, if b_{n_1,n_2} and c_{n_1,n_2} are the matrices of Fourier coefficients of $\phi|T_n$ and $\phi|S_n$, respectively, then, since we are assuming that $a_{1,1} = 1$, the eigenvalues of T_n and S_n on ϕ are, respectively, $b_{1,1}$ and $c_{1,1}$. Now, by (9.8-9), we have:

$$b_{1,1} = na_{n,1} \tag{9.10}$$

$$c_{1,1} = na_{1,n} \tag{9.11}$$

Thus:

$$\phi|T_n = na_{n,1}\phi \tag{9.12}$$

$$\phi|S_n = na_{1,n}\phi \tag{9.13}$$

The Euler product (9.1) now follows by applying the operator in (9.6) to $\tilde{\phi}$, and comparing the eigenvalues on both sides.

Now, by (9.12-13) and (9.8-9), we have:

$$a_{n,1}a_{n_1,n_2} = \sum_{\substack{p_0p_1p_2=n \\ p_1|n_1 \\ p_2|n_2}} a_{\frac{n_1p_0}{p_1}, \frac{n_2p_1}{p_2}} \tag{9.14}$$

$$a_{1,n}a_{n_1,n_2} = \sum_{\substack{p_0p_1p_2=n \\ p_0|n_1 \\ p_1|n_2}} a_{\frac{n_1p_1}{p_0},\frac{n_2p_2}{p_1}} \tag{9.15}$$

EXERCISE: Give another proof of (9.1) based on these formulae.

As a special case of either of these formulae, we have:

$$a_{n_1,1}a_{1,n_2} = \sum_{d|(n_1,n_2)} a_{\frac{n_1}{d},\frac{n_2}{d}} \tag{9.16}$$

This implies the double L-series formula (9.2).

We may now prove the multiplicativity (9.3). Indeed, if $n_2 = n_2' = 1$, this follows from (9.12), and from the corresponding multiplicativity assertion for the operators T_n (cf. [30], 3.2.1); similarly, if $n_1 = n_1' = 1$, this follows from the corresponding multiplicativity for the S_n. The general case follows from these two special cases using (9.16) and Möbius inversion.

Let us prove (9.5). This, too, we will deduce from the double L-series formula (9.16).

We will require the following formula, valid for x and y sufficiently small:

$$\sum_{k_1=0}^{\infty}\sum_{k_2=0}^{\infty} S_{k_1,k_2}(a,b,c)x^{k_1}y^{k_2} =$$

$$(1-abcxy)(1-ay)^{-1}(1-by)^{-1}(1-cy)^{-1}(1-bcx)^{-1}(1-cax)^{-1}(1-abx)^{-1} \tag{9.17}$$

To prove this, we have:

$$S_{k_1,k_2}(a,b,c) = (b-a)^{-1}(c-b)^{-1}(a-c)^{-1}\{a^{k_1+k_2+2}(b^{k_1+1}-c^{k_1+1}) + b^{k_1+k_2+2}(c^{k_1+1}-a^{k_1+1}) + c^{k_1+k_2+2}(a^{k_1+1}-b^{k_1+1})\}$$

Summing the geometric series:

$$\sum_{k_1=0}^{\infty}\sum_{k_2=0}^{\infty} S_{k_1,k_2}(a,b,c)x^{k_1}y^{k_2} = (b-a)^{-1}(c-b)^{-1}(a-c)^{-1}\cdot [(1-ay)^{-1}(a^2b(1-abx)^{-1}-a^2c(1-acx)^{-1}) + (1-by)^{-1}(b^2c(1-bcx)^{-1}-b^2a(1-bax)^{-1}) + (1-cy)^{-1}(c^2a(1-cax)^{-1}-c^2b(1-cbx)^{-1})].$$

After some algebra, we obtain (9.17).

Now, we have, after (9.4):

$$1 - a_{1,p}x + a_{p,1}x^2 - x^3 = (1-\alpha x)(1-\beta x)(1-\gamma x)$$

$$1 - a_{p,1}y + a_{1,p}y^2 - y^3 = (1-\beta\gamma y)(1-\gamma\alpha y)(1-\alpha\beta y) .$$

Here, of course, $\alpha\beta\gamma = 1$. Now, by (9.16), we have:

$$\sum_{k_1=0}^{\infty} \sum_{k_2=0}^{\infty} a_{p^{k_1},p^{k_2}} x^{k_1} y^{k_2}$$

$$= (1-xy) \sum_{k_1=0}^{\infty} \sum_{k_2=0}^{\infty} \sum_{k \leq \mathrm{Min}(k_1,k_2)} a_{p^{k_1-k},p^{k_2-k}} x^{k_1} y^{k_2}$$

$$= (1-xy) \sum_{k_1=0}^{\infty} \sum_{k_2=0}^{\infty} a_{p^{k_1},1}\, a_{1,p^{k_2}} x^{k_1} y^{k_2}$$

$$= (1-xy)(1-a_{p,1}y+a_{1,p}y^2-y^3)^{-1}(1-a_{1,p}x+a_{p,1}x^2-x^3)^{-1}$$

$$= (1-xy)(1-\alpha y)^{-1}(1-\beta y)^{-1}(1-\gamma y)^{-1}(1-\beta\gamma x)^{-1}(1-\gamma\alpha x)^{-1}(1-\alpha\beta x)^{-1}$$

$$= \sum_{k_1=0}^{\infty} \sum_{k_2=0}^{\infty} S_{k_1,k_2}(\alpha,\beta,\gamma) x^{k_1} y^{k_2} .$$

Comparing coefficients on both sides, we obtain (9.5).

Let us finally prove (9.8-9), whose proofs we deferred. Actually, since S_n is related to T_n by the involution ι, (9.9) follows from (9.8). We shall prove (9.8) only. T_n is the sum of all double cosets $\Gamma\alpha\Gamma$ with α of determinant n. A complete set of right coset representatives for these double cosets is the set of all:

$$\begin{pmatrix} p_2 & q_2 & q_3 \\ & p_1 & q_1 \\ & & p_0 \end{pmatrix}$$

where p_0, p_1, p_2 are positive integers, $p_0 p_1 p_2 = n$, and q_1, q_3 both run through a complete set of residue classes mod p_0, and q_2 runs through a complete set of residue classes mod p_1. Thus:

$$|n_1 n_2|^{-1} b_{n_1,n_2} W\left(\begin{pmatrix} n_1 n_2 & & \\ & n_1 & \\ & & 1 \end{pmatrix}\tau\right)$$

$$= \frac{1}{n^3} \sum_{p_0 p_1 p_2 = n} \sum_{\substack{q_1, q_3 \bmod p_0 \\ q_2 \bmod p_1}} \int_0^n \int_0^n \int_0^n$$

$$\phi\left(\begin{pmatrix} p_2 & q_2 & q_3 \\ & p_1 & q_1 \\ & & p_0 \end{pmatrix}\begin{pmatrix} 1 & \xi_2 & \xi_3 \\ & 1 & \xi_1 \\ & & 1 \end{pmatrix}\tau\right) e(-n_1\xi_1 - n_2\xi_2)\, d\xi_3\, d\xi_2\, d\xi_1$$

Now, we have:

$$\begin{pmatrix} p_2 & q_2 & q_3 \\ & p_1 & q_1 \\ & & p_0 \end{pmatrix}\begin{pmatrix} 1 & \xi_2 & \xi_3 \\ & 1 & \xi_1 \\ & & 1 \end{pmatrix} = \begin{pmatrix} 1 & \eta_2 & \eta_3 \\ & 1 & \eta_1 \\ & & 1 \end{pmatrix}\begin{pmatrix} p_2 & & \\ & p_1 & \\ & & p_0 \end{pmatrix}$$

where:

$$\eta_1 = \frac{1}{p_0}(p_1\xi_1 + q_1)$$

$$\eta_2 = \frac{1}{p_1}(p_2\xi_2 + q_2)$$

$$\eta_3 = \frac{1}{p_0}(p_2\xi_3 + q_2\xi_1 + q_3)$$

Our previous integral thus equals:

$$\frac{1}{n^3}\sum_{p_0p_1p_2=n}\frac{p_0}{p_1}\frac{p_1}{p_2}\frac{p_0}{p_2}\sum_{\substack{q_1,q_3 \bmod p_0\\ q_2 \bmod p_1}} e\left(\frac{q_1n_1}{p_1}+\frac{q_2n_2}{p_2}\right).$$

$$\int_{q_1/p_0}^{p_1^2p_2+\frac{q_1}{p_0}}\int_{q_2/p_1}^{p_2^2\,p_0+\frac{q_2}{p_1}}\int_{L(\eta_1)}^{p_2^2p_1+L(\eta_1)}$$

$$\phi\left(\begin{pmatrix}1 & \eta_2 & \eta_3\\ & 1 & \eta_1\\ & & 1\end{pmatrix}\begin{pmatrix}p_2 & & \\ & p_1 & \\ & & p_0\end{pmatrix}\tau\right) e\left(-\frac{n_1p_0}{p_1}\eta_1-\frac{n_2p_1}{p_2}\eta_2\right)d\eta_3d\eta_2d\eta_1$$

where:

$$L(\eta_1) = \frac{q_2}{p_0}\eta_1 - \frac{q_1q_2-p_1q_3}{p_0p_1}$$

Using the periodicity of the integral, this is the same as:

$$\frac{1}{n^3}\sum_{p_0p_1p_2=n}\frac{p_0^2}{p_2^2}\sum_{\substack{q_1,q_3 \bmod p_0\\ q_2 \bmod p_1}} e\left(\frac{q_1n_1}{p_1}+\frac{q_2n_2}{p_2}\right).$$

$$\int_0^{p_1^2p_2}\int_0^{p_2^2p_0}\int_0^{p_2^2p_1}\phi\left(\begin{pmatrix}1 & \eta_2 & \eta_3\\ & 1 & \eta_1\\ & & 1\end{pmatrix}\begin{pmatrix}p_2 & & \\ & p_1 & \\ & & p_0\end{pmatrix}\tau\right)\cdot e\left(-\frac{n_1p_0}{p_1}\eta_1-\frac{n_2p_1}{p_2}\eta_2\right)d\eta_3d\eta_2d\eta_1$$

The integral vanishes unless $p_1|p_0n_1, p_2|p_1n_2$. If this is the case, we have:

$$\sum_{\substack{q_1,q_3 \bmod p_0 \\ q_2 \bmod p_1}} e\left(\frac{q_1n_1}{p_1}+\frac{q_2n_2}{p_2}\right) = \begin{cases} p_0^2p_1 & \text{if } p_1|n_1,\ p_2|n_2 \\ 0 & \text{otherwise.} \end{cases}$$

Thus, our previous expression equals:

$$\frac{1}{n^3} \sum_{\substack{p_0p_1p_2=n \\ p_1|n_1 \\ p_2|n_2}} \frac{p_0^2}{p_2^2}\cdot p_0^2p_1\cdot p_1^2p_2\cdot p_2^2p_0\cdot p_2^2p_1\cdot$$

$$\left|\frac{n_1p_0}{p_1}\cdot\frac{n_2p_1}{p_2}\right|^{-1} a_{\frac{n_1p_0}{p_1},\frac{n_2p_1}{p_2}} W\left(\begin{pmatrix} n_1n_2 & & \\ & n_1 & \\ & & 1\end{pmatrix}\tau\right)$$

$$= |n_1n_2|^{-1}\cdot n \sum_{\substack{p_0p_1p_2=n \\ p_1|n_1 \\ p_2|n_2}} a_{\frac{n_1p_0}{p_1},\frac{n_2p_1}{p_2}} W\left(\begin{pmatrix} n_1n_2 & & \\ & n_1 & \\ & & 1\end{pmatrix}\tau\right).$$

(9.8) now follows.

CHAPTER X

THE MELLIN TRANSFORMS OF THE WHITTAKER FUNCTIONS

We will prove in this chapter, that, denoting $W(\tau) = W^{(\nu_1,\nu_2)}_{1,1}(\tau, w_1)$, we have:

$$\int_0^\infty \int_0^\infty W\begin{pmatrix} y_1 y_2 & & \\ & y_1 & \\ & & 1 \end{pmatrix} y_1^{s_1-1} y_2^{s_2-1} \frac{dy_1}{y_1}\frac{dy_2}{y_2} = \tag{10.1}$$

$$\frac{1}{4}\pi^{-s_1-s_2} \frac{\Gamma\left(\frac{s_1+\alpha}{2}\right)\Gamma\left(\frac{s_1+\beta}{2}\right)\Gamma\left(\frac{s_1+\gamma}{2}\right)\Gamma\left(\frac{s_2-\alpha}{2}\right)\Gamma\left(\frac{s_2-\beta}{2}\right)\Gamma\left(\frac{s_2-\gamma}{2}\right)}{\Gamma\left(\frac{s_1+s_2}{2}\right)}$$

for sufficiently large s_1, s_2, where:

$$\alpha = -\nu_1 - 2\nu_2 + 1$$

$$\beta = -\nu_1 + \nu_2$$

$$\gamma = 2\nu_1 + \nu_2 - 1$$

By the Mellin inversion formula, this implies that:

$$W\begin{pmatrix} y_1 y_2 & & \\ & y_1 & \\ & & 1 \end{pmatrix} =$$

$$\frac{1}{4}\frac{1}{(2\pi i)^2} \int_{\sigma-i\infty}^{\sigma+i\infty}\int_{\sigma-i\infty}^{\sigma+i\infty} \frac{\Gamma\left(\frac{s_1+\alpha}{2}\right)\Gamma\left(\frac{s_1+\beta}{2}\right)\Gamma\left(\frac{s_1+\gamma}{2}\right)\Gamma\left(\frac{s_2-\alpha}{2}\right)\Gamma\left(\frac{s_2-\beta}{2}\right)\Gamma\left(\frac{s_2-\gamma}{2}\right)}{\Gamma\left(\frac{s_1+s_2}{2}\right)} \cdot$$

$$(\pi y_1)^{1-s_1}(\pi y_2)^{1-s_2} ds_1 ds_2 \tag{10.2}$$

where σ is sufficiently large.

(10.1) appears to be a simple problem in advanced calculus -- the evaluation of a multiple integral, if one begins with the definition (3.11) of $W(\tau)$. One should be able to do this integral (in five variables). However, despite our best efforts, a direct approach failed to yield (10.1). The method by which we prove this is indirect -- it consists of evaluating the Mellin transform of the Eisenstein series in two different ways, firstly, using the series definition of the Eisenstein series, and, secondly, using the Fourier expansions of Chapter VII. The two expressions for this Mellin transform are not valid for the same values of the parameters ν_1, ν_2, s_1, s_2, but after analytic continuation, they may be compared (using the functional equation of the Riemann zeta function). There results a simpler integral -- a triple integral, which is still very difficult, but which we shall be able to evaluate.

We have not actually worked out all the technical difficulties caused by the degenerate terms of the Eisenstein series. However, there is no doubt that our calculations are correct (the constant $\frac{1}{4}$ could be wrong).

Further comments on the problem of the Mellin transforms were made in Chapter VIII, where may be found another example of the trick of evaluating a transform of the Eisenstein series to obtain information about a transform of the Whittaker functions. Yet another example is given by Friedberg [3].

Since the proof of (10.1) is very difficult, it may be instructive to first consider the corresponding problem for GL(2). We will devote the first part of this chapter to the Mellin transforms of the GL(2) Whittaker functions.

During our discussion of the GL(2) case, $\mathcal{H}$ will denote the usual complex upper half-plane, and $\tau = x+iy$ will denote a point in $\mathcal{H}$. Let $\nu \in \mathbb{C}$, and denote, as in Chapter I:

$$I_\nu(\tau) = y^\nu$$
$$I_\nu^*(\tau) = \pi^{-\nu}\Gamma(\nu)I_\nu(\tau)$$

Also as in Chapter I, we have the Whittaker function, defined by

$$W_\nu(\tau) = \int_{-\infty}^{\infty} I_\nu^*\left(\begin{pmatrix} & 1 \\ -1 & \end{pmatrix}\begin{pmatrix} 1 & x \\ & 1 \end{pmatrix}\tau\right) e(-x)dx$$

for $re(\nu) > \frac{1}{2}$, and having analytic continuation to all values of ν. That is:

$$W_\nu(iy) = \pi^{-\nu}\Gamma(\nu)\int_{-\infty}^{\infty} \left(\frac{y}{x^2+y^2}\right)^\nu e(-x)dx \tag{10.3}$$

If $re(s) > re(\nu) > \frac{1}{2}$, the following integral is convergent:

$$M_\nu(s) = \int_0^\infty W_\nu(iy)y^s \frac{dy}{y}$$

We give two proofs of the Mellin transform formula:

$$M_\nu(s) = \tfrac{1}{2}\pi^{-\frac{1}{2}-s}\,\Gamma\left(\frac{s+\nu}{2}\right)\Gamma\left(\frac{s+1-\nu}{2}\right) \tag{10.4}$$

Both proofs will depend on the following integral formula (valid if $re(s) > 1+re(\nu)$):

$$\int_0^\infty \frac{y^{\nu+s}}{(y^2+1)^s}\frac{dy}{y} = \tfrac{1}{2}\frac{\Gamma\left(\frac{s+\nu}{2}\right)\Gamma\left(\frac{s-\nu}{2}\right)}{\Gamma(s)} \tag{10.5}$$

We will evaluate a similar integral for GL(3) in this chapter, and yet another in Bump and Goldfeld [2]. Thus, even though (10.5) may be deduced from standard integrals, it may be of interest to consider a proof which is similar to proofs of these GL(3) integrals.

Indeed, we have:

$$\Gamma(s)\int_0^\infty \frac{y^{\nu+s}}{(y^2+1)^s}\frac{dy}{y} = \int_0^\infty\int_0^\infty e^{-t(y^2+1)y^{-1}}y^{\nu}t^{s}\frac{dy}{y}\frac{dt}{t}\,.$$

Substituting $u = ty$, $v = t/y$, this equals:

$$\tfrac{1}{2}\int_0^\infty\int_0^\infty e^{-u-v}u^{\frac{s+\nu}{2}}v^{\frac{s-\nu}{2}}\frac{du}{u}\frac{dv}{v}$$

The variables now separate, whence (10.5).

We now give the first proof of (10.4). By (10.3), we have:

$$\begin{aligned} W_\nu(iy) &= \int_{-\infty}^\infty\int_0^\infty e^{-t}\left(\frac{ty}{\pi(x^2+y^2)}\right)^{\nu} e(-x)\,\frac{dt}{t}\,dx \\ &= \int_{-\infty}^\infty\int_0^\infty e^{-\pi(x^2+y^2)ty^{-1}}t^{\nu}e(-x)\,\frac{dt}{t}\,dx \end{aligned}$$

As:

$$\int_{-\infty}^{\infty} e^{-\pi x^2 t y^{-1}} e(-x)dx = \sqrt{\frac{y}{t}}\, e^{-\pi y t^{-1}}$$

we have:

$$W_\nu(iy) = \sqrt{y}\int_0^\infty e^{-\pi y(t+t^{-1})} t^{\nu-\frac{1}{2}} \frac{dt}{t}$$

From this we have the analytic continuation of the Whittaker function, and the functional equation (1.7). Also, in terms of the usual K-Bessel function, we have by Watson [38], section 6.22 (15):

$$W_\nu(iy) = 2\sqrt{y}\, K_{\nu-\frac{1}{2}}(2\pi y)$$

(cf. (1.5)). Furthermore, we now may compute $M_\nu(s)$. Indeed:

$$\int_0^\infty W_\nu(iy) y^s \frac{dy}{y} = \int_0^\infty\int_0^\infty e^{-\pi y(t+t^{-1})} t^{\nu-\frac{1}{2}} y^{s+\frac{1}{2}} \frac{dt}{t}\frac{dy}{y}$$

$$= \int_0^\infty\int_0^\infty e^{-y} y^{s+\frac{1}{2}} (\pi(t+t^{-1}))^{-s-\frac{1}{2}} t^{\nu-\frac{1}{2}} \frac{dt}{t}\frac{dy}{y}$$

$$= \pi^{-s-\frac{1}{2}}\Gamma(s+\tfrac{1}{2})\int_0^\infty \frac{t^{s+\nu}}{(1+t^2)^{s+\frac{1}{2}}}\frac{dt}{t} .$$

(10.4) now follows from (10.5).

We now turn to the second proof of (10.4). This proof is indirect, depending on the Fourier expansion of the Eisenstein series. Although more difficult than the previous proof for GL(2), this proof has the advantage that it generalizes to GL(3) to yield (10.1). We have been unable to give a direct proof of that formula.

We shall prove that $M_\nu(s)$ has meromorphic continuation in ν and s, and if $re(\nu)-1 > re(s) > 0$, we will show that:

$$M_\nu(s) = \pi^{-s-\frac{1}{2}} \frac{\Gamma(\nu)\Gamma\left(\frac{s+1-\nu}{2}\right)}{\Gamma\left(\frac{\nu-s}{2}\right)} \int_0^\infty \frac{y^{\nu+s}}{(y^2+1)^\nu} \frac{dy}{y} \tag{10.6}$$

Thus (10.4) follows from (10.5) (interchanging the roles of s and ν! Is this role reversal significant??).

Naively, one could attempt to argue as follows: by (10.3), we have:

$$\int_0^\infty W_\nu(iy)y^s \frac{dy}{y} = \pi^{-\nu}\Gamma(\nu) \int_0^\infty \int_{-\infty}^\infty \left(\frac{y}{x^2+y^2}\right)^\nu y^s e(-x)dx \frac{dy}{y} .$$

We would like to interchange the order of integration, then, substituting xy for y, we obtain:

$$\pi^{-\nu}\Gamma(\nu) \int_{-\infty}^\infty x^{s-\nu}e(-x)dx \cdot \int_0^\infty \frac{y^{\nu+s}}{(y^2+1)^\nu} \frac{dy}{y} .$$

Resulting is an expression involving the integral of (10.6). Unfortunately, interchanging the order of integration is not justifiable. Nor is it easy to attach a meaning to the divergent integral:

$$\int_{-\infty}^{\infty} x^{s-\nu} e(-x)\,dx$$

Finally, and fatally, the integrals:

$$\int_0^{\infty} W_\nu(iy)y^s \frac{dy}{y} \qquad \text{and} \qquad \int_0^{\infty} \frac{y^{\nu+s}}{(y^2+1)^\nu} \frac{dy}{y}$$

do not converge for the same values of s and ν -- ever! Thus, we may not prove (10.6) by a straightforward transformation of the integral. To overcome these difficulties, we turn to the theory of the Eisenstein series.

As in Chapter I, let:

$$G_\nu(\tau) = \tfrac{1}{2}\pi^{-\nu}\Gamma(\nu) \sum{}' \frac{y^\nu}{|m\tau+n|^2} \tag{10.7}$$

Here the summation is over all nonzero pairs $(m,n) \in \mathbb{Z}^2$. Then:

$$G_\nu(iy) = \lambda(2\nu)y^\nu + \lambda(2-2\nu)y^{1-\nu} + 2\sum_{n=1}^{\infty} |n|^{-\nu}\sigma_{2\nu-1}(|n|)W(niy)$$

where:

$$\lambda(s) = \pi^{-\frac{s}{2}}\,\Gamma(\tfrac{s}{2})\zeta(s)$$

Also, let $\hat{G}_\nu(\tau)$ be the function obtained from $G_\nu(\tau)$ by subtracting the constant term:

$$\hat{G}_\nu(\tau) = G_\nu(\tau) - \lambda(2\nu)y^\nu - \lambda(2-2\nu)y^{1-\nu}$$

Let $re(s) > re(\nu) > 1$. Using the fact that $G_\nu(iy) = G_\nu(i/y)$, we may compute the Mellin transform as follows:

$$2\zeta(s+\nu)\zeta(s+1-\nu)M_\nu(s) = \int_0^\infty \hat{G}_\nu(iy)y^s \frac{dy}{y} = \int_1^\infty \hat{G}_\nu(iy)(y^s+y^{-s}) \frac{dy}{y}$$

$$+ \lambda(2\nu)\{\int_1^\infty y^{\nu-s} \frac{dy}{y} - \int_0^1 y^{s+\nu} \frac{dy}{y}\}$$

$$+ \lambda(2-2\nu)\{\int_1^\infty y^{1-\nu-s} \frac{dy}{y} - \int_0^1 y^{s+1-\nu} \frac{dy}{y}\}$$

$$= \int_1^\infty \hat{G}_\nu(iy)(y^s+y^{-s}) \frac{dy}{y} - \lambda(2\nu)\{\frac{1}{\nu+s} + \frac{1}{\nu-s}\} - \lambda(2-2\nu)\{\frac{1}{1-\nu+s} + \frac{1}{1-\nu-s}\} . \tag{10.8}$$

This expression gives the analytic continuation of $M_\nu(s)$ to values of s and ν no longer required to satisfy $re(s) > re(\nu)$. On the contrary, let us evaluate (10.8) in the case $re(\nu)-1 > re(s) > 0$. Noting that:

$$\int_1^\infty y^{1-\nu}(y^s+y^{-s}) \frac{dy}{y} = -\frac{1}{1-\nu+s} - \frac{1}{1-\nu-s}$$

we may rewrite (10.7) as:

$$\int_1^\infty (G_\nu(iy) - \lambda(2\nu)y^\nu)(y^s + y^{-s})\,\frac{dy}{y} - \lambda(2\nu)\{\frac{1}{\nu+s} + \frac{1}{\nu-s}\}$$

Let:

$$\phi_\nu(y) = \begin{cases} \lambda(2\nu)y^\nu & \text{if} \quad y > 1; \\ \lambda(2\nu)y^{-\nu} & \text{if} \quad y < 1, \end{cases}$$

$$\psi_\nu(y) = \begin{cases} \lambda(2\nu)y^{-\nu} & \text{if} \quad y > 1; \\ \lambda(2\nu)y^\nu & \text{if} \quad y < 1. \end{cases}$$

We may rewrite our previous expression as:

$$\int_0^\infty (G_\nu(iy) - \phi_\nu(y))y^s\,\frac{dy}{y} - \lambda(2\nu)\{\frac{1}{\nu+s} + \frac{1}{\nu-s}\}$$

Now let us consider the incomplete Eisenstein series:

$$g_\nu(\tau) = \tfrac{1}{2}\pi^{-\nu}\Gamma(\nu) \sum_{\substack{m\neq 0\\ n\neq 0}} \frac{y^\nu}{|m\tau+n|^{2\nu}}$$

We have:

$$G_\nu(iy) = g_\nu(iy) + \lambda(2\nu)y^\nu + \lambda(2\nu)y^{-\nu}\ .$$

This follows from (10.7). Indeed, the first term on the right is the sum of all terms in (10.7) with $m,n \neq 0$; the second is the sum of the terms with $m = 0,\ n \neq 0$; and the third is the sum of all terms with $m \neq 0,\ n = 0$. Rewriting this:

$$G_\nu(iy) = g_\nu(iy) + \phi_\nu(y) + \psi_\nu(y)$$

and noting that:

$$\int_0^\infty \psi_\nu(y) y^s \frac{dy}{y} = \lambda(2\nu)\{\frac{1}{\nu+s} + \frac{1}{\nu-s}\}$$

our previous expression becomes, simply:

$$\int_0^\infty g_\nu(iy) y^s \frac{dy}{y}$$

Noting that:

$$\int_0^\infty \frac{y^{\nu+s}}{(m^2y^2+n^2)^{2\nu}} = n^{-\nu+s} m^{-\nu-s} \int_0^\infty \frac{y^{\nu+s}}{(y^2+1)^\nu} \frac{dy}{y}$$

we obtain:

$$2\zeta(s+\nu)\zeta(s+1-\nu)M_\nu(s) = 2\zeta(\nu+s)\zeta(\nu-s)\pi^{-\nu}\Gamma(\nu)\int_0^\infty \frac{y^{\nu+s}}{(y^2+1)^\nu}\frac{dy}{y}.$$

(10.6), and hence (10.4) now follows from the functional equation of the Riemann zeta function!

We turn now to the $GL(3)$ case, and the proof of (10.1). Here, unfortunately, we have not worked out all the details arising from the constant terms in the Fourier expansions of the Eisenstein series. However, there can be no doubt that this computation is correct. Let:

$$M_{(\nu_1,\nu_2)}(s_1,s_2) = \int_0^\infty\int_0^\infty W\begin{pmatrix} y_1y_2 & & \\ & y_1 & \\ & & 1\end{pmatrix} y_1^{s_1-1}y_2^{s_2-1}\frac{dy_1}{y_1}\frac{dy_2}{y_2}$$

defined initially for $re(s_1)$, $re(s_2)$ large in comparison with $re(\nu_1)$, $re(\nu_2)$. We will prove that this Mellin transform has meromorphic continuation, and, if $re(\nu_1)$, $re(\nu_2)$ are large, that:

$$M_{(\nu_1,\nu_2)}(s_1,s_2) = \pi^{-s_1-s_2-\frac{1}{2}}\Gamma\left(\frac{3\nu_1}{2}\right)\Gamma\left(\frac{3\nu_2}{2}\right)\Gamma\left(\frac{3\nu_1+3\nu_2-1}{2}\right)\Gamma\left(\frac{s_1+\alpha}{2}\right)\Gamma\left(\frac{s_2-\gamma}{2}\right)$$
$$\Gamma\left(\frac{1-s_1-\alpha}{2}\right)^{-1}\Gamma\left(\frac{1-s_2+\gamma}{2}\right)^{-1}.$$
$$\int_0^\infty\int_0^\infty\int_{-\infty}^\infty (x^2+y_1^2+y_1^2y_2^2)^{-\frac{3\nu_1}{2}}((1-x)^2+y_2^2+y_1^2y_2^2)^{-\frac{3\nu_2}{2}}y_1^{s_1+\gamma}y_2^{s_2-\alpha}\frac{dy_1}{y_1}\frac{dy_2}{y_2} \tag{10.9}$$

We will prove this as follows. Let $G_{(\nu_1,\nu_2)}$ be the Eisenstein series of Chapter VII. Let:

$$H_{(\nu_1,\nu_2)}(\tau) = \int_0^1 G_{(\nu_1,\nu_2)}\left(\begin{pmatrix} 1 & & x \\ & 1 & \\ & & 1 \end{pmatrix}\tau\right)dx .$$

The method of proof of (10.9) is the same as the second proof of (10.9)-- we will evaluate the Mellin transform of $H_{(\nu_1,\nu_2)}$ in two different ways. Firstly, if s_1, s_2 are large in comparison with ν_1, ν_2, we will evaluate the Mellin transform by means of the Fourier expansion; and if ν_1, ν_2 are large, by means of the series (7.4). The two resulting expressions are compared, after analytic continuation, yielding (10.9).

It is necessary to modify $H_{(\nu_1,\nu_2)}$ before the Mellin transform can converge. Suppose first that s_1 and s_2 are large. By (4.1), (4.2) and (4.7):

$$H_{(\nu_1,\nu_2)}(\tau) = \sum_{n_1,n_2} \phi_{n_1,n_2}(\tau) .$$

Let us consider just the sum of the nondegenerate terms:

$$\hat{H}_{(\nu_1,\nu_2)}(\tau) = \sum_{n_1,n_2\neq 0} \phi_{n_1,n_2}(\tau)$$

we will show that:

$$\int_0^\infty\int_0^\infty \hat{H}_{(\nu_1,\nu_2)}\begin{pmatrix} y_1y_2 & & \\ & y_1 & \\ & & 1 \end{pmatrix} y_1^{s_1-1} y_2^{s_2-1} \frac{dy_1}{y_1}\frac{dy_2}{y_2}$$

$$= 4\zeta(s_1+\alpha)\zeta(s_1+\beta)\zeta(s_1+\gamma)\zeta(s_2-\alpha)\zeta(s_2-\beta)\zeta(s_2-\gamma)\zeta(s_1+s_2)^{-1}M_{(\nu_1,\nu_2)}(s_1,s_2)$$

(10.10)

On the other hand, if ν_1, ν_2 are large, we consider the incomplete Eisenstein series (compare (7.4)):

$$g_{(\nu_1,\nu_2)}(\tau) = \tfrac{1}{4}\pi^{-3\nu_1-3\nu_2+\frac{1}{2}}\Gamma\left(\frac{3\nu_1}{2}\right)\Gamma\left(\frac{3\nu_2}{2}\right)\Gamma\left(\frac{3\nu_1+3\nu_2-1}{2}\right)\zeta(3\nu_1+3\nu_2-1)$$

$$I_{(\nu_1,\nu_2)}(\tau) \qquad \sum_{\substack{A_1C_2+B_1B_2+C_1A_2=0 \\ A_1,B_1 \neq 0 \\ A_2,B_2 \neq 0}} [(A_1x_3+B_1x_1+C_1)^2+(A_1x_2+B_1)^2y_1^2+A_1^2y_1^2y_2^2]^{-\frac{3\nu_1}{2}}$$

$$[(A_2x_4-B_2x_2+C_2)^2+(A_2x_1-B_2)^2y_2^2+A_2^2y_1^2y_2^2]^{-\frac{3\nu_2}{2}}$$

Let:

$$h_{(\nu_1,\nu_2)}(\tau) = \int_0^1 g_{(\nu_1,\nu_2)}\left(\begin{pmatrix} 1 & & x \\ & 1 & \\ & & 1 \end{pmatrix}\tau\right)dx \ .$$

We will show that:

$$\int_0^\infty\int_0^\infty h_{(\nu_1,\nu_2)}\begin{pmatrix} y_1y_2 & & \\ & y_1 & \\ & & 1\end{pmatrix} y_1^{s_1-1}y_2^{s_2-1}\,\frac{dy_1}{y_1}\,\frac{dy_2}{y_2}$$

$$= 4\pi^{-3\nu_1-3\nu_2+\frac{1}{2}}\Gamma\left(\frac{3\nu_1}{2}\right)\Gamma\left(\frac{3\nu_2}{2}\right)\Gamma\left(\frac{3\nu_1+3\nu_2-1}{2}\right).$$

$$\zeta(1-s_1-\alpha)\zeta(s_1+\beta)\zeta(s_1+\gamma)\zeta(s_2-\alpha)\zeta(s_2-\beta)\zeta(1-s_2+\gamma)\zeta(s_1+s_2)^{-1}.$$

$$\int_0^\infty\int_0^\infty\int_{-\infty}^\infty (x^2+y_1^2+y_1^2y_2^2)^{-\frac{3\nu_1}{2}}((1-x)^2+y_2^2+y_1^2y_2^2)^{-\frac{3\nu_2}{2}}y_1^{s_1+\gamma}y_2^{s_2-\alpha}\,\frac{dy_1}{y_1}\,\frac{dy_2}{y_2} \tag{10.11}$$

That the integrals (10.10) and (10.11) are equal, after analytic continuation, we will not prove, as we have not worked out the technical difficulties. However, this is unquestionably true. (10.9) follows by the functional equation of the Riemann zeta function.

Let us prove (10.10). By Theorem 7.2, the left-hand side equals:

$$\sum_{n_1,n_2\neq 0} a_{n_1,n_2}|n_1|^{-s_1}|n_2|^{-s_2}M_{(\nu_1,\nu_2)}(s_1,s_2) .$$

By (4.13), this equals:

$$4\sum_{n_1=1}^{\infty}\sum_{n_2=1}^{\infty} a_{n_1,n_2}n_1^{-s_1}n_2^{-s_2}M_{(\nu_1,\nu_2)}(s_1,s_2) .$$

(10.10) now follows from (8.11) and (9.2).

Let us prove (10.11). We have:

$$h_{(\nu_1,\nu_2)}\begin{pmatrix} y_1y_2 & & \\ & y_1 & \\ & & 1\end{pmatrix} = \tfrac{1}{4}\pi^{-3\nu_1-3\nu_2+\frac{1}{2}}\Gamma\left(\frac{3\nu_1}{2}\right)\Gamma\left(\frac{3\nu_2}{2}\right)\Gamma\left(\frac{3\nu_1+3\nu_2-1}{2}\right)\zeta(3\nu_1+3\nu_2-1).$$

$$y_1^{2\nu_1+\nu_2}\,y_2^{\nu_1+2\nu_2}\sum_{\substack{A_1,B_1\neq 0\\ A_2,B_2\neq 0}}\ \sum_{\substack{C_1 \bmod A_1\\ C_2 \bmod A_2\\ A_1C_2+B_1B_2+C_1A_2=0}}$$

$$\int_{-\infty}^{\infty}[(A_1x_3+C_1)^2+B_1^2y_1^2+A_1^2y_1^2y_2^2]^{-\frac{3\nu_1}{2}}$$

$$[(-A_2x_3+C_2)^2+B_2^2y_2^2+A_2^2y_1^2y_2^2]^{-\frac{3\nu_2}{2}}\,dx_3$$

The substitution:

$$x = A_2B_1^{-1}B_2^{-2}(A_1x_3+C_1)$$

$$1-x = A_1B_1^{-1}B_2^{-1}(-A_2x_3+C_2)$$

shows that the integral is independent of C_1, C_2. The number of C_1 mod A_1, C_2 mod A_2 such that $A_1C_2 + B_1B_2 + C_1A_2 = 0$ is (A_1,A_2) if $(A_1,A_2)|B_1B_2$; zero otherwise. Also, let us note that the term corresponding to A_1,A_2,B_1,B_2 has the same contribution as the term corresponding to $|A_1|,|A_2|,|B_1|,|B_2|$. Thus:

$$h_{(\nu_1,\nu_2)}\begin{pmatrix} y_1y_2 & & \\ & y_1 & \\ & & 1 \end{pmatrix}$$

$$= 4\pi^{-3\nu_1-3\nu_2+\frac{1}{2}}\Gamma\left(\frac{3\nu_1}{2}\right)\Gamma\left(\frac{3\nu_2}{2}\right)\Gamma\left(\frac{3\nu_1+3\nu_2-1}{2}\right)\zeta(3\nu_1+3\nu_2-1)y_1^{2\nu_1+\nu_2}y_2^{\nu_1+2\nu_2}.$$

$$\sum_{A_1,A_2=1}^{\infty}(A_1,A_2)\sum_{\substack{B_1,B_2=1\\(A_1,A_2)|B_1B_2}}^{\infty} B_1B_2A_1^{-1}A_2^{-1}.$$

$$\int_{-\infty}^{\infty}[(B_1B_2A_2^{-1}x)^2 + B_1^2y_1^2 + A_1^2y_1^2y_2^2]^{-\frac{3\nu_1}{2}}$$

$$[(B_1B_2A_1^{-1}(1-x))^2 + B_2^2y_2^2 + A_2^2y_1^2y_2^2]^{-\frac{3\nu_2}{2}}\,dx \qquad (10.12)$$

Substituting this into the left-hand side of (10.11), interchanging the integral with the summation, and replacing y_1,y_2 by $B_2A_2^{-1}y_1$, $B_1A_1^{-1}y_2$

respectively, we see that the left-hand side of (10.11) equals:

$$4\pi^{-3\nu_1-3\nu_2+\frac{1}{2}}\Gamma\left(\frac{3\nu_1}{2}\right)\Gamma\left(\frac{3\nu_2}{2}\right)\Gamma\left(\frac{3\nu_1+3\nu_2-1}{2}\right)\zeta(3\nu_1+3\nu_2-1)\cdot$$

$$\left\{\sum_{A_1,A_2=1}^{\infty}\ \sum_{\substack{B_1B_2=1\\(A_1,A_2)|B_1B_2}}^{\infty}(A_1,A_2)A_1^{\beta-s_2}A_2^{-\beta-s_1}B_1^{s_2-\gamma-1}B_2^{s_1+\alpha-1}\right\}\cdot$$

$$\int_0^\infty\int_0^\infty\int_{-\infty}^\infty (x^2+y_1^2+y_1^2y_2^2)^{-\frac{3\nu_1}{2}}((1-x)^2+y_2^2+y_1^2y_2^2)^{-\frac{3\nu_1}{2}}y_1^{s_1+\gamma}y_2^{s_2-\alpha}\frac{dy_1}{y_1}\frac{dy_2}{y_2}.$$

(10.11) will follow from the following identity:

$$\sum_{A_1,A_2=1}^{\infty}\ \sum_{\substack{B_1,B_2=1\\(A_1,A_2)|B_1B_2}}(A_1,A_2)A_1^{\beta-s_2}A_2^{-\beta-s_1}B_1^{s_2-\gamma-1}B_2^{s_1+\alpha-1}$$

$$= \zeta(1-s_1-\alpha)\zeta(s_1+\beta)\zeta(s_1+\gamma)\zeta(s_2-\alpha)\zeta(s_2-\beta)\zeta(1-s_2+\gamma)\zeta(s_1+s_2)^{-1}\zeta(3\nu_1+3\nu_2-1)^{-1}$$

(10.13)

To prove this, note that the left side has an Euler product; the p-factor equals:

$$\sum_{k_1=0}^{\infty}\sum_{k_2=0}^{\infty}\sum_{\substack{h_1,h_2=0\\ \min(k_1,k_2)\le h_1+h_2}}^{\infty} p^{\min(k_1,k_2)}u^{k_1}v^{k_2}x^{h_1}y^{h_2} \qquad (10.14)$$

where:

$$u = p^{\beta-s_2}$$

$$v = p^{-\beta-s_1}$$

$$x = p^{s_2-\gamma-1}$$

$$y = p^{s_1+\alpha-1}$$

We leave it to the reader to show that (10.14) equals:

$$(1-u)^{-1}(1-v)^{-1}(1-x)^{-1}(1-y)^{-1}(1-uvpx)^{-1}(1-uvpy)^{-1}(1-uv)(1-uvpxy) \ .$$

Taking the product over all primes, we obtain (10.13). (10.11) and hence (10.9) now follow.

(10.1) will follow from (10.9) when we prove the integral formula:

$$\int_0^\infty\int_0^\infty (x^2+y_1^2+y_1^2y_2^2)^{-\frac{3\nu_1}{2}}((1-x)^2+y_2^2+y_1^2y_2^2)^{-\frac{3\nu_2}{2}} y_1^{s_1+\gamma} y_2^{s_2-\alpha} \frac{dy_1}{y_1}\frac{dy_2}{y_2}$$

$$= \frac{\sqrt{\pi}}{4}\Gamma\left(\frac{3\nu_1}{2}\right)^{-1}\Gamma\left(\frac{3\nu_2}{2}\right)^{-1}\Gamma\left(\frac{3\nu_1+3\nu_2-1}{2}\right)^{-1}\Gamma\left(\frac{1-\alpha-s_1}{2}\right)\Gamma\left(\frac{s_1+\beta}{2}\right)\Gamma\left(\frac{s_1+\gamma}{2}\right)\Gamma\left(\frac{s_2-\alpha}{2}\right)$$

$$\Gamma\left(\frac{s_2-\beta}{2}\right)\Gamma\left(\frac{1-s_2+\gamma}{2}\right)\Gamma\left(\frac{s_1+s_2}{2}\right)^{-1} \ . \qquad (10.15)$$

Indeed, we have:

$$\Gamma\left(\frac{3\nu_1}{2}\right)\Gamma\left(\frac{3\nu_2}{2}\right)\int_0^\infty\int_0^\infty (x^2+y_1^2+y_1^2y_2^2)^{-\frac{3\nu_1}{2}}((1-x)^2+y_2^2+y_1^2y_2^2)^{-\frac{3\nu_2}{2}} y_1^{s_1+\gamma} y_2^{s_2-\alpha}\frac{dy_1}{y_1}\frac{dy_2}{y_2}$$

$$= \int_{-\infty}^\infty\int_0^\infty\int_0^\infty\int_0^\infty\int_0^\infty e^{-t_1(x^2+y_1^2+y_1^2y_2^2) - t_2((1-x)^2+y_2^2+y_1^2y_2^2)}\cdot$$

$$y_1^{s_1+\gamma} y_2^{s_2-\alpha} t_1^{\frac{3\nu_1}{2}} t_2^{\frac{3\nu_2}{2}} \frac{dt_1}{t_1}\frac{dt_2}{t_2}\frac{dy_1}{y_1}\frac{dy_2}{y_2}\,dx\ .$$

Let us make the substitution:

$$U = y_1^2t_1 \qquad V = y_1^2y_2^2t_1 \qquad W = y_2^2t_2 \qquad X = y_1^2y_2^2t_2\ .$$

Our previous integral equals:

$$\tfrac{1}{4}\int_{-\infty}^\infty\int_0^\infty\int_0^\infty\int_0^\infty\int_0^\infty e^{-UWX^{-1}x^2-UWV^{-1}(1-x)^2}e^{-U-V-W-X}$$

$$U^{a+\frac{1}{2}}V^{b-\frac{1}{2}}W^{c+\frac{1}{2}}X^{d-\frac{1}{2}}\frac{dU}{U}\frac{dV}{V}\frac{dW}{W}\frac{dX}{X}\,dx$$

where:

$$a = \tfrac{1}{2}(2\nu_1+\nu_2-s_2) = \tfrac{1}{2}(1+\gamma-s_2)$$

$$b = \tfrac{1}{2}(\nu_1-\nu_2+s_2) = \tfrac{1}{2}(s_2-\beta)$$

$$c = \tfrac{1}{2}(\nu_1+2\nu_2-s_1) = \tfrac{1}{2}(1-\alpha-s_1)$$

$$d = \tfrac{1}{2}(-\nu_1+\nu_2+s_1) = \tfrac{1}{2}(s_1+\beta)\ .$$

By the integral formula:

$$\int_{-\infty}^{\infty} e^{-Ax^2-B(1-x)^2}dx = \sqrt{\frac{\pi}{A+B}}\, e^{-AB(A+B)^{-1}}$$

our previous integral equals:

$$\frac{\sqrt{\pi}}{4}\int_0^{\infty}\int_0^{\infty}\int_0^{\infty}\int_0^{\infty} e^{-u-v-w-X}(X+V)^{-1}U^aV^bW^cX^d e^{-UW(X+V)^{-1}} \frac{dU}{U}\frac{dV}{V}\frac{dW}{W}\frac{dX}{X}\,.$$

Recall that (cf. Gradshteyn and Ryzhik [7]):

$$\int_0^{\infty} e^{-A^2t-B^2t^{-1}}t^{\nu}\frac{dt}{t} = 2\left|\frac{B}{A}\right|^{\nu}K_{\nu}(2|AB|) \tag{10.16}$$

Thus, if $Y = X+V$, we have, in terms of Euler's beta function:

$$\int_0^{\infty}\int_0^{\infty} e^{-V-X}(X+V)^{-1}V^bX^d e^{-UW(X+V)^{-1}}\frac{dV}{V}\frac{dX}{X}$$

$$= \int_0^{\infty} e^{-Y-UWY^{-1}}Y^{-\frac{1}{2}}\int_0^{Y}(Y-X)^{b-1}X^{d-1}dXdY$$

$$= B(b,d)\int_0^{\infty} e^{-Y-UWY^{-1}}Y^{b+d-\frac{1}{2}}\frac{dY}{Y}$$

$$= 2B(b,d)(UW)^{\frac{1}{2}(b+d-\frac{1}{2})}K_{b+d-\frac{1}{2}}(2\sqrt{UW})$$

Our previous integral thus equals:

$$\frac{\sqrt{\pi}}{2} B(b,d) \int_0^\infty \int_0^\infty e^{-U-W} U^{a+\frac{1}{2}(b+d-\frac{1}{2})} W^{c+\frac{1}{2}(b+d-\frac{1}{2})} K_{b+d-\frac{1}{2}}(2\sqrt{UW}) \frac{dU}{U} \frac{dW}{W}$$

Now let $u = \sqrt{UW}$, $v = \sqrt{U/W}$. We obtain:

$$\sqrt{\pi} B(b,d) \int_0^\infty \int_0^\infty e^{-uv-uv^{-1}} u^{\frac{1}{2}(3\nu_1+3\nu_2-1)} v^{\frac{1}{2}(\nu_1-\nu_2+s_1-s_2)} K_{\frac{1}{2}(s_1+s_2-1)}(2u) \frac{du}{u} \frac{dv}{v}$$

$$= 2\sqrt{\pi} B(b,d) \int_0^\infty u^{\frac{1}{2}(3\nu_1+3\nu_2-1)} K_{\frac{1}{2}(s_1+s_2-1)}(2u) K_{\frac{1}{2}(\nu_1-\nu_2+s_1-s_2)}(2u) \frac{du}{u} .$$

(10.15), and hence (10.1), now follow from Gradshteyn and Ryzhik [7], 6.576.4.

REFERENCES

[1] W. BAILY, Introductory Lectures on Automorphic Forms, Iwanami Shoten and Princeton University Press (1973).

[2] D. BUMP AND D. GOLDFELD, A Kronecker Limit Formula for Cubic Fields, to appear (1983).

[2b] J. DIEUDONNÉ AND J. CARRELL, Invariant Theory, Academic Press (1971).

[3] S. FRIEDBERG, A Global Approach to the Rankin-Selberg Convolution for GL(3, ℤ), to appear.

[4] S. GELBART, Automorphic Forms on Adele Groups, Princeton University Press, Annals of Mathematics Study #83 (1975).

[5] R. GODEMENT AND H. JACQUET, Zeta Functions of Simple Algebras, Springer Verlag, Lecture Notes in Mathematics #260 (1972).

[6] R. GOODMAN AND N. WALLACH, Whittaker Vectors and Conical Vectors, J. Funct. Anal. 39 #2 (1980), 199-279.

[7] I. GRADSHTEYN AND I. RYSHIK, Tables of Integrals, Series and Products. Corrected and enlarged edition. Academic Press (1980).

[8] HARISH-CHANDRA, Automorphic Forms on Semisimple Lie Groups, Springer Verlag, Lecture Notes in Mathematics #62.

[9] S. HELGASON, Differential Geometry and Symmetric Spaces, Academic Press (1962).

[10] J. HUMPHREYS, Introduction to Lie Algebras and Representation Theory, Springer Verlag (1972).

[11] K. IMAI AND A. TERRAS, The Fourier Expansions of Eisenstein Series for GL(3, ℤ), Trans. AMS 273 (1982), #2, 679-694.

[12] H. JACQUET, Dirichlet Series for the Group GL(n), in Automorphic Forms, Representation Theory and Arithmetic, Springer Verlag and the Tata Institute (1981).

[13] H. JACQUET, Fonctions de Whittaker associees aux Groups de Chevalley, Bull. Soc. Math. France 95 (1967), 243-309.

[14] H. JACQUET AND R. LANGLANDS, Automorphic Forms on GL(2), Springer Verlag, Lecture Notes in Mathematics #114 (1970).

[15] H. JACQUET, I.I. PIATETSKI-SHAPIRO AND J. SHALIKA, Automorphic Forms on GL(3), Part I and II, Annals of Math. 109 (1979), 169-258.

[16] H. JACQUET, I.I. PIATETSKI-SHAPIRO AND J. SHALIKA, Rankin-Selberg Convolutions, American J. Math. 105 (1982) #2, 367-464.

[17] B. KOSTANT, On Whittaker Vectors and Representation Theory, Inventiones Math. 48 (1978), 101-184.

[18] T. KUBOTA, Elementary Theory of Eisenstein Series, Kodansha Ltd. and John Wiley and Sons (1973).

[19] S. LANG, SL(2,ℝ), Addison Wesley (1975).

[20] R. LANGLANDS, On the Functional Equations Satisfied by Eisenstein Series (1964), finally published as Springer Verlag, Lecture Notes in Mathematics #544 (1976).

[21] R. LANGLANDS, Problems in the Theory of Automorphic Forms, in Lectures in Modern Analysis and Applications, Springer Verlag, Lecture Notes in Mathematics #170 (1970), 18-86.

[22] H. MAASS, Über eine Neue Art von Nichtanalytischen Automorphen Funktionen und die Bestimmung Dirichletscher Reihen durch Funktional Gleichungen, Math. Annalen 121 (1949), 141-183.

[22b] H. MAASS, Siegel's Modular Forms and Dirichlet Series, Springer Verlag, Lecture Notes in Mathematics #216 (1971).

[22c] I. MACDONALD, Symmetric Functions and Hall Polynomials, Oxford (1979).

[23] R. NARASIMHAN, Several Complex Variables, University of Chicago (1971).

[24] I.I. PIATETSKI-SHAPIRO, Euler Subgroups, in Lie Groups and their Representations, John Wiley and Sons (1975), 597-620.

[25] I.I. PIATETSKI-SHAPIRO, Multiplicity One Theorems, in Automorphic Forms, Representations, and L-Functions, Proceedings of Symposia in Pure Mathematics #XXXII (A. Borel, Ed.), Part II, 209-212.

[26] PROSKURIN, Expansions of Automorphic Functions, Proc. Steklov. Inst. Math. 116 (1982), 119-141. In Russian.

[27] S. RAMANUJAN, On certain trigonometric sums and their applications in the theory of numbers, Trans. Cambridge Phil. Soc. 22 (1918), #13, 259-276. Reprinted in Ramanujan's Collected Works, Cambridge University Press (1927) #21, (now available from Chelsea).

[28] J. SHALIKA, The Multiplicity One Theorem for GL(n), Annals of Math. 100 (1974), 171-193.

[29] G. SHIFFMANN, Integrals d'entrelacement et Fonctions de Whittaker, Bull. Soc. Math. France 99 (1971), 3-72.

[30] G. SHIMURA, Introduction to the Arithmetic Theory of Automorphic Forms, Iwanami Shoten and Princeton University Press (1971).

[31] T. SHINTANI, On an explicit Formula for Class-1 "Whittaker Functions" on GL_n over P-adic Fields, Proc. Japan Acad. 52 (1976), 180-182.

[32] TAMAGAWA, On the Zeta Functions of a Division Algebra, Ann. of Math. 77 (1963), 387-405.

[33] A. TERRAS, Harmonic Analysis on Symmetric Spaces, Lecture notes in perpetual revision, UCSD, to appear.

[34] A. TERRAS, On Automorphic Forms for the General Linear Group, Rocky Mountain J. Math. 12 (1982).

[35] A. TERRAS, The Chowla-Selberg Method for Fourier Expansion of Higher Rank Eisenstein Series, to appear.

[36] V. VARADARAJAN, Lie Groups Lie Algebras and their Representations, Prentice Hall (1974).

[37] I. VINOGRADOV AND L. TAKHTADZHYAN, Theory of Eisenstein Series for the Group SL $(3,\mathbb{R})$ and its application to a binary problem, J. Sov. Math. 18 (1982), #3, 293-324.

[38] G. WATSON, Bessel Functions, Cambridge University Press (1922).

[39] H. WEYL, The Classical Groups, Princeton University Press (1939).

[40] E. WHITTAKER AND G. WATSON, A Course of Modern Analysis, Cambridge University Press, fourth edition (1927).

Vol. 926: Geometric Techniques in Gauge Theories. Proceedings, 1981. Edited by R. Martini and E.M.de Jager. IX, 219 pages. 1982.

Vol. 927: Y. Z. Flicker, The Trace Formula and Base Change for GL (3). XII, 204 pages. 1982.

Vol. 928: Probability Measures on Groups. Proceedings 1981. Edited by H. Heyer. X, 477 pages. 1982.

Vol. 929: Ecole d'Eté de Probabilités de Saint-Flour X - 1980. Proceedings, 1980. Edited by P.L. Hennequin. X, 313 pages. 1982.

Vol. 930: P. Berthelot, L. Breen, et W. Messing, Théorie de Dieudonné Cristalline II. XI, 261 pages. 1982.

Vol. 931: D.M. Arnold, Finite Rank Torsion Free Abelian Groups and Rings. VII, 191 pages. 1982.

Vol. 932: Analytic Theory of Continued Fractions. Proceedings, 1981. Edited by W.B. Jones, W.J. Thron, and H. Waadeland. VI, 240 pages. 1982.

Vol. 933: Lie Algebras and Related Topics. Proceedings, 1981. Edited by D. Winter. VI, 236 pages. 1982.

Vol. 934: M. Sakai, Quadrature Domains. IV, 133 pages. 1982.

Vol. 935: R. Sot, Simple Morphisms in Algebraic Geometry. IV, 146 pages. 1982.

Vol. 936: S.M. Khaleelulla, Counterexamples in Topological Vector Spaces. XXI, 179 pages. 1982.

Vol. 937: E. Combet, Intégrales Exponentielles. VIII, 114 pages. 1982.

Vol. 938: Number Theory. Proceedings, 1981. Edited by K. Alladi. IX, 177 pages. 1982.

Vol. 939: Martingale Theory in Harmonic Analysis and Banach Spaces. Proceedings, 1981. Edited by J.-A. Chao and W.A. Woyczyński. VIII, 225 pages. 1982.

Vol. 940: S. Shelah, Proper Forcing. XXIX, 496 pages. 1982.

Vol. 941: A. Legrand, Homotopie des Espaces de Sections. VII, 132 pages. 1982.

Vol. 942: Theory and Applications of Singular Perturbations. Proceedings, 1981. Edited by W. Eckhaus and E.M. de Jager. V, 363 pages. 1982.

Vol. 943: V. Ancona, G. Tomassini, Modifications Analytiques. IV, 120 pages. 1982.

Vol. 944: Representations of Algebras. Workshop Proceedings, 1980. Edited by M. Auslander and E. Lluis. V, 258 pages. 1982.

Vol. 945: Measure Theory. Oberwolfach 1981, Proceedings. Edited by D. Kölzow and D. Maharam-Stone. XV, 431 pages. 1982.

Vol. 946: N. Spaltenstein, Classes Unipotentes et Sous-groupes de Borel. IX, 259 pages. 1982.

Vol. 947: Algebraic Threefolds. Proceedings, 1981. Edited by A. Conte. VII, 315 pages. 1982.

Vol. 948: Functional Analysis. Proceedings, 1981. Edited by D. Butković, H. Kraljević, and S. Kurepa. X, 239 pages. 1982.

Vol. 949: Harmonic Maps. Proceedings, 1980. Edited by R.J. Knill, M. Kalka and H.C.J. Sealey. V, 158 pages. 1982.

Vol. 950: Complex Analysis. Proceedings, 1980. Edited by J. Eells. IV, 428 pages. 1982.

Vol. 951: Advances in Non-Commutative Ring Theory. Proceedings, 1981. Edited by P.J. Fleury. V, 142 pages. 1982.

Vol. 952: Combinatorial Mathematics IX. Proceedings, 1981. Edited by E. Billington, S. Oates-Williams, and A.P. Street. XI, 443 pages. 1982.

Vol. 953: Iterative Solution of Nonlinear Systems of Equations. Proceedings, 1982. Edited by R. Ansorge, Th. Meis, and W. Törnig. VII, 202 pages. 1982.

Vol. 954: S.G. Pandit, S.G. Deo, Differential Systems Involving Impulses. VII, 102 pages. 1982.

Vol. 955: G. Gierz, Bundles of Topological Vector Spaces and Their Duality. IV, 296 pages. 1982.

Vol. 956: Group Actions and Vector Fields. Proceedings, 1981. Edited by J.B. Carrell. V, 144 pages. 1982.

Vol. 957: Differential Equations. Proceedings, 1981. Edited by D.G. de Figueiredo. VIII, 301 pages. 1982.

Vol. 958: F.R. Beyl, J. Tappe, Group Extensions, Representations, and the Schur Multiplicator. IV, 278 pages. 1982.

Vol. 959: Géométrie Algébrique Réelle et Formes Quadratiques, Proceedings, 1981. Edité par J.-L. Colliot-Thélène, M. Coste, L. Mahé, et M.-F. Roy. X, 458 pages. 1982.

Vol. 960: Multigrid Methods. Proceedings, 1981. Edited by W. Hackbusch and U. Trottenberg. VII, 652 pages. 1982.

Vol. 961: Algebraic Geometry. Proceedings, 1981. Edited by J.M. Aroca, R. Buchweitz, M. Giusti, and M. Merle. X, 500 pages. 1982.

Vol. 962: Category Theory. Proceedings, 1981. Edited by K.H. Kamps, D. Pumplün, and W. Tholen, XV, 322 pages. 1982.

Vol. 963: R. Nottrot, Optimal Processes on Manifolds. VI, 124 pages. 1982.

Vol. 964: Ordinary and Partial Differential Equations. Proceedings, 1982. Edited by W.N. Everitt and B.D. Sleeman. XVIII, 726 pages. 1982.

Vol. 965: Topics in Numerical Analysis. Proceedings, 1981. Edited by P.R. Turner. IX, 202 pages. 1982.

Vol. 966: Algebraic K-Theory. Proceedings, 1980, Part I. Edited by R.K. Dennis. VIII, 407 pages. 1982.

Vol. 967: Algebraic K-Theory. Proceedings, 1980. Part II. VIII, 409 pages. 1982.

Vol. 968: Numerical Integration of Differential Equations and Large Linear Systems. Proceedings, 1980. Edited by J. Hinze. VI, 412 pages. 1982.

Vol. 969: Combinatorial Theory. Proceedings, 1982. Edited by D. Jungnickel and K. Vedder. V, 326 pages. 1982.

Vol. 970: Twistor Geometry and Non-Linear Systems. Proceedings, 1980. Edited by H.-D. Doebner and T.D. Palev. V, 216 pages. 1982.

Vol. 971: Kleinian Groups and Related Topics. Proceedings, 1981. Edited by D.M. Gallo and R.M. Porter. V, 117 pages. 1983.

Vol. 972: Nonlinear Filtering and Stochastic Control. Proceedings, 1981. Edited by S.K. Mitter and A. Moro. VIII, 297 pages. 1983.

Vol. 973: Matrix Pencils. Proceedings, 1982. Edited by B. Kågström and A. Ruhe. XI, 293 pages. 1983.

Vol. 974: A. Draux, Polynômes Orthogonaux Formels - Applications. VI, 625 pages. 1983.

Vol. 975: Radical Banach Algebras and Automatic Continuity. Proceedings, 1981. Edited by J.M. Bachar, W.G. Bade, P.C. Curtis Jr., H.G. Dales and M.P. Thomas. VIII, 470 pages. 1983.

Vol. 976: X. Fernique, P.W. Millar, D.W. Stroock, M. Weber, Ecole d'Eté de Probabilités de Saint-Flour XI - 1981. Edited by P.L. Hennequin. XI, 465 pages. 1983.

Vol. 977: T. Parthasarathy, On Global Univalence Theorems. VIII, 106 pages. 1983.

Vol. 978: J. Ławrynowicz, J. Krzyż, Quasiconformal Mappings in the Plane. VI, 177 pages. 1983.

Vol. 979: Mathematical Theories of Optimization. Proceedings, 1981. Edited by J.P. Cecconi and T. Zolezzi. V, 268 pages. 1983.

Vol. 980: L. Breen. Fonctions thêta et théorème du cube. XIII, 115 pages. 1983.

Vol. 981: Value Distribution Theory. Proceedings, 1981. Edited by I. Laine and S. Rickman. VIII, 245 pages. 1983.

Vol. 982: Stability Problems for Stochastic Models. Proceedings, 1982. Edited by V. V. Kalashnikov and V. M. Zolotarev. XVII, 295 pages. 1983.

Vol. 983: Nonstandard Analysis-Recent Developments. Edited by A.E. Hurd. V, 213 pages. 1983.

Vol. 984: A. Bove, J. E. Lewis, C. Parenti, Propagation of Singularities for Fuchsian Operators. IV, 161 pages. 1983.

Vol. 985: Asymptotic Analysis II. Edited by F. Verhulst. VI, 497 pages. 1983.

Vol. 986: Séminaire de Probabilités XVII 1981/82. Proceedings. Edited by J. Azéma and M. Yor. V, 512 pages. 1983.

Vol. 987: C. J. Bushnell, A. Fröhlich, Gauss Sums and p-adic Division Algebras. XI, 187 pages. 1983.

Vol. 988: J. Schwermer, Kohomologie arithmetisch definierter Gruppen und Eisensteinreihen. III, 170 pages. 1983.

Vol. 989: A. B. Mingarelli, Volterra-Stieltjes Integral Equations and Generalized Ordinary Differential Expressions. XIV, 318 pages. 1983.

Vol. 990: Probability in Banach Spaces IV. Proceedings, 1982. Edited by A. Beck and K. Jacobs. V, 234 pages. 1983.

Vol. 991: Banach Space Theory and its Applications. Proceedings, 1981. Edited by A. Pietsch, N. Popa and I. Singer. X, 302 pages. 1983.

Vol. 992: Harmonic Analysis, Proceedings, 1982. Edited by G. Mauceri, F. Ricci and G. Weiss. X, 449 pages. 1983.

Vol. 993: R. D. Bourgin, Geometric Aspects of Convex Sets with the Radon-Nikodým Property. XII, 474 pages. 1983.

Vol. 994: J.-L. Journé, Calderón-Zygmund Operators, Pseudo-Differential Operators and the Cauchy Integral of Calderón. VI, 129 pages. 1983.

Vol. 995: Banach Spaces, Harmonic Analysis, and Probability Theory. Proceedings, 1980–1981. Edited by R.C. Blei and S.J. Sidney. V, 173 pages. 1983.

Vol. 996: Invariant Theory. Proceedings, 1982. Edited by F. Gherardelli. V, 159 pages. 1983.

Vol. 997: Algebraic Geometry – Open Problems. Edited by C. Ciliberto, F. Ghione and F. Orecchia. VIII, 411 pages. 1983.

Vol. 998: Recent Developments in the Algebraic, Analytical, and Topological Theory of Semigroups. Proceedings, 1981. Edited by K.H. Hofmann, H. Jürgensen and H. J. Weinert. VI, 486 pages. 1983.

Vol. 999: C. Preston, Iterates of Maps on an Interval. VII, 205 pages. 1983.

Vol. 1000: H. Hopf, Differential Geometry in the Large, VII, 184 pages. 1983.

Vol. 1001: D.A. Hejhal, The Selberg Trace Formula for PSL(2, ℝ). Volume 2. VIII, 806 pages. 1983.

Vol. 1002: A. Edrei, E.B. Saff, R.S. Varga, Zeros of Sections of Power Series. VIII, 115 pages. 1983.

Vol. 1003: J. Schmets, Spaces of Vector-Valued Continuous Functions. VI, 117 pages. 1983.

Vol. 1004: Universal Algebra and Lattice Theory. Proceedings, 1982. Edited by R.S. Freese and O.C. Garcia. VI, 308 pages. 1983.

Vol. 1005: Numerical Methods. Proceedings, 1982. Edited by V. Pereyra and A. Reinoza. V, 296 pages. 1983.

Vol. 1006: Abelian Group Theory. Proceedings, 1982/83. Edited by R. Göbel, L. Lady and A. Mader. XVI, 771 pages. 1983.

Vol. 1007: Geometric Dynamics. Proceedings, 1981. Edited by J. Palis Jr. IX, 827 pages. 1983.

Vol. 1008: Algebraic Geometry. Proceedings, 1981. Edited by J. Dolgachev. V, 138 pages. 1983.

Vol. 1009: T. A. Chapman, Controlled Simple Homotopy Theory and Applications. III, 94 pages. 1983.

Vol. 1010: J.-E. Dies, Chaînes de Markov sur les permutations. IX, 226 pages. 1983.

Vol. 1011: J.M. Sigal. Scattering Theory for Many-Body Quantum Mechanical Systems. IV, 132 pages. 1983.

Vol. 1012: S. Kantorovitz, Spectral Theory of Banach Space Operators. V, 179 pages. 1983.

Vol. 1013: Complex Analysis – Fifth Romanian-Finnish Seminar. Part 1. Proceedings, 1981. Edited by C. Andreian Cazacu, N. Boboc, M. Jurchescu and I. Suciu. XX, 393 pages. 1983.

Vol. 1014: Complex Analysis – Fifth Romanian-Finnish Seminar. Part 2. Proceedings, 1981. Edited by C. Andreian Cazacu, N. Boboc, M. Jurchescu and I. Suciu. XX, 334 pages. 1983.

Vol. 1015: Equations différentielles et systèmes de Pfaff dans le champ complexe – II. Seminar. Edited by R. Gérard et J.P. Ramis. V, 411 pages. 1983.

Vol. 1016: Algebraic Geometry. Proceedings, 1982. Edited by M. Raynaud and T. Shioda. VIII, 528 pages. 1983.

Vol. 1017: Equadiff 82. Proceedings, 1982. Edited by H. W. Knobloch and K. Schmitt. XXIII, 666 pages. 1983.

Vol. 1018: Graph Theory, Łagów 1981. Proceedings, 1981. Edited by M. Borowiecki, J. W. Kennedy and M. M. Sysło. X, 289 pages. 1983.

Vol. 1019: Cabal Seminar 79–81. Proceedings, 1979–81. Edited by A. S. Kechris, D. A. Martin and Y. N. Moschovakis. V, 284 pages. 1983.

Vol. 1020: Non Commutative Harmonic Analysis and Lie Groups. Proceedings, 1982. Edited by J. Carmona and M. Vergne. V, 187 pages. 1983.

Vol. 1021: Probability Theory and Mathematical Statistics. Proceedings, 1982. Edited by K. Itô and J.V. Prokhorov. VIII, 747 pages. 1983.

Vol. 1022: G. Gentili, S. Salamon and J.-P. Vigué. Geometry Seminar "Luigi Bianchi", 1982. Edited by E. Vesentini. VI, 177 pages. 1983.

Vol. 1023: S. McAdam, Asymptotic Prime Divisors. IX, 118 pages. 1983.

Vol. 1024: Lie Group Representations I. Proceedings, 1982–1983. Edited by R. Herb, R. Lipsman and J. Rosenberg. IX, 369 pages. 1983.

Vol. 1025: D. Tanré, Homotopie Rationnelle: Modèles de Chen, Quillen, Sullivan. X, 211 pages. 1983.

Vol. 1026: W. Plesken, Group Rings of Finite Groups Over p-adic Integers. V, 151 pages. 1983.

Vol. 1027: M. Hasumi, Hardy Classes on Infinitely Connected Riemann Surfaces. XII, 280 pages. 1983.

Vol. 1028: Séminaire d'Analyse P. Lelong – P. Dolbeault – H. Skoda. Années 1981/1983. Edité par P. Lelong, P. Dolbeault et H. Skoda. VIII, 328 pages. 1983.

Vol. 1029: Séminaire d'Algèbre Paul Dubreil et Marie-Paule Malliavin. Proceedings, 1982. Edité par M.-P. Malliavin. V, 339 pages. 1983.

Vol. 1030: U. Christian, Selberg's Zeta-, L-, and Eisensteinseries. XII, 196 pages. 1983.

Vol. 1031: Dynamics and Processes. Proceedings, 1981. Edited by Ph. Blanchard and L. Streit. IX, 213 pages. 1983.

Vol. 1032: Ordinary Differential Equations and Operators. Proceedings, 1982. Edited by W. N. Everitt and R. T. Lewis. XV, 521 pages. 1983.

Vol. 1033: Measure Theory and its Applications. Proceedings, 1982. Edited by J. M. Belley, J. Dubois and P. Morales. XV, 317 pages. 1983.

Vol. 1034: J. Musielak, Orlicz Spaces and Modular Spaces. V, 222 pages. 1983.